Digital Innovations in Architecture, Engineering and Construction

The Architecture, Engineering and Construction (AEC) industry is experiencing an unprecedented transformation from conventional labor-intensive activities to automation using innovative digital technologies and processes. This new paradigm also requires systemic changes focused on social, economic and sustainability aspects. Within the scope of Industry 4.0, digital technologies are a key factor in interconnecting information between the physical built environment and the digital virtual ecosystem. The most advanced virtual ecosystems allow to simulate the built to enable a real-time data-driven decision-making. This Book Series promotes and expedites the dissemination of recent research, advances, and applications in the field of digital innovations in the AEC industry. Topics of interest include but are not limited to:

- Industrialization: digital fabrication, modularization, cobotics, lean.
- Material innovations: bio-inspired, nano and recycled materials.
- Reality capture: computer vision, photogrammetry, laser scanning, drones.
- Extended reality: augmented, virtual and mixed reality.
- Sustainability and circular building economy.
- Interoperability: building/city information modeling.
- Interactive and adaptive architecture.
- Computational design: data-driven, generative and performance-based design.
- Simulation and analysis: digital twins, virtual cities.
- Data analytics: artificial intelligence, machine/deep learning.
- Health and safety: mobile and wearable devices, QR codes, RFID.
- Big data: GIS, IoT, sensors, cloud computing.
- Smart transactions, cybersecurity, gamification, blockchain.
- Quality and project management, business models, legal prospective.
- Risk and disaster management.

Katia Perini · Francesca Mosca

Nature-Based Cities

Performance-Driven Design Approaches for Climate Change Adaptation

Springer

Katia Perini
Dipartimento di Architettura e Design
Università di Genova
Genoa, Italy

Francesca Mosca
Dipartimento di Architettura e Design
Università di Genova
Genoa, Italy

ISSN 2731-7269 ISSN 2731-7277 (electronic)
Digital Innovations in Architecture, Engineering and Construction
ISBN 978-3-031-94611-0 ISBN 978-3-031-94612-7 (eBook)
https://doi.org/10.1007/978-3-031-94612-7

This Springer imprint is published by the registered company Springer Nature Switzerland AG
The registered company address is: Gewerbestrasse 11, 6330 Cham, Switzerland

Competing Interests The authors have no competing interests to declare that are relevant to the content of this manuscript.

Contents

About the Authors

Katia Perini is Associate Professor at the Architecture and Design Department, Polytechnic School of the University of Genoa (Italy). Katia Perini main research interests are effects and performances of nature-based solutions in the field of environmental and economic sustainability in (of) urban areas and building/urban design. Katia Perini obtained the EU Ph.D. label in 2012 at the University of Genoa. She was a visiting student at the Delft University of Technology and visiting scholar at Columbia University (NY, USA, Fulbright grant) and at the Technische Universität München (TUM, DAAD award). Katia Perini is PI and local PI of several research projects, e.g., ECOLOPES (FET-H2020, ecolopes.org), URBAN GENERATION (PRIN-PNRR), BIG4LIFE (LIFE22). Over 90 publications (peer-reviewed journals, books, etc.). https://orcid.org/0000-0003-0415-8246.

Francesca Mosca is a post-doctoral researcher at the Architecture and Design Department, Polytechnic School of the University of Genoa (Italy). She currently contributes to the URBAN GENERATION (PRIN-PNRR) project. She obtained her EU Ph.D. label in May 2025 at the University of Genoa in the framework of the FET-H2020 ECOLOPES project. She was visiting Ph.D. student at the Technische Universität München (TUM) where she joined the teaching team of the Green Technologies for Landscape Architecture (GTLA) research group. Her main research interests include the assessment of nature-based solutions performances in relation to environmental and social sustainability and the identification of climate adaptation strategies for urban areas. https://orcid.org/0000-0001-8898-8727.

Abbreviations

A-UHI	Atmospheric Urban Heat Island
CAD	Computer-Aided Design
CBD	Convention on Biological Diversity
CC	Climate Change
CCA	Climate Change Adaptation
CFD	Computational Fluid Dynamics
CL-UHI	Canopy Layer Urban Heat Island
COP	Conference of Parties
DBT	Dry Bulb Temperature
DEM	Digital Elevation Model
DR	Draught Rate
EPA	Environmental Protection Agency
EPW	EnergyPlus Weather File
GBI	Green Blue Infrastructure
GBIF	Global Biodiversity Information Facility
GH	Grasshopper
GHG	Greenhouse Gases
GI	Green Infrastructure
IPCC	Intergovernmental Panel for Climate Change
IUCN	International Union for Conservation of Nature
KPI	Key Performance Indicator
LAD	Leaf Area Density
LAI	Leaf Area Index
LCZ	Local Climate Zone
LID	Low Impact Development
LU	Land Use
MCDM	Multi-criteria Decision Making
MOO	Multi-objective Optimization
MRT	Mean Radiant Temperature
NbS	Nature-Based Solutions
PET	Physiological Equivalent Temperature

PM	Particulate Matter
PMV	Predicted Mean Vote
PPD	Predicted Percentage of Dissatisfied
PUC	Piano Urbanistico Comunale
RH	Relative Humidity
SC	Sponge City
SDGs	Sustainable Development Goals
SET	Standard Effective Temperature
SUDS	Sustainable Urban Drainage System
S-UHI	Surface Urban Heat Island
SWMM	Storm Water Management Model
UBL-UHI	Urban Boundary Layer Urban Heat Island
UHI	Urban Heat Island
UN	United Nations
UNEP	United Nations Environmental Protection
UNFCCC	United Nations Framework Convention on Climate Change
UNU-EHS	United Nations University—Environment and Human Security
UTCI	Universal Thermal Climate Index
WSUD	Water Sensitive Urban Design

Chapter 1
The Introduction to Nature-Based Cities

Abstract This chapter introduces the primary challenges urban areas are currently facing, regarding both environmental and ecological issues. It outlines the impacts of climate change (CC) on cities and their inhabitants and provides examples of nature-based adaptation solutions for the building scale and for open spaces. These topics are addressed in connection to the most relevant Sustainable Development Goal and other international environmental goals, from a design and planning perspective. Moreover, the chapter presents the structure of the book Nature-Based Cities, which aims to address the rising urban complexity by proposing approaches, methodologies, and tools to design Nature-based Solutions (NbS) for climate change adaptation.

Keywords Nature-based solutions · Computational design · Climate change adaptation · Microclimate · Water management · Urban areas · Climate change impacts

1.1 Ecological and Environmental Challenges of Urbanization

More than half of the world's population nowadays lives in urban areas, a trend that is projected to increase in the next decades (United Nations, Department of Economic and Social Affairs, Population Division 2018). Urbanization poses significant challenges in terms of ecosystem health, human wellbeing, and environmental quality (Benedict and McMahon 2001; Nelson et al. 2013; McDonald et al. 2020). Poor air quality (European Environment Agency 2022), heatwaves exacerbated by the urban heat island effect (Taha 2008; Chen et al. 2023), and rainwater management (United Nations 2012; Jamali et al. 2020) are among the most relevant environmental issues faced by urban areas and their inhabitants. Additionally, urbanization has severe impacts on surrounding rural areas and at the global scale, for example, due to the rising extraction and use of raw materials (Lederer et al. 2021) and the production of greenhouse gases (GHG) emissions (Du and Xia 2018). Finally, the increase in frequency and intensity of extreme events caused by GHG emissions and

K. Perini and F. Mosca, *Nature-Based Cities*, Digital Innovations in Architecture, Engineering and Construction, https://doi.org/10.1007/978-3-031-94612-7_1

other anthropic activities poses significant challenges in terms of urban resilience (Beheshtian et al. 2018).

Cities and their surroundings are urban ecosystems, i.e., complex socio-ecological systems where human populations live, comprising built infrastructure as well as physical and biological components (Maes et al. 2016). They can be fragmented, densely populated and characterized by a high level of anthropic pressure, lack of green spaces, habitat fragmentation and local biodiversity loss (Grimm et al. 2000; Wu 2008), with a strong impact on their liveability (Aronson et al. 2017).

Land use change and the reduction of natural surfaces, besides being responsible for biodiversity loss, are among the most relevant causes of environmental degradation in cities (Groffman et al. 2017; McDonald et al. 2020; Rashid et al. 2022). These processes result in an increasing separation of people from nature and the reduction of ecosystem services provision, fundamental for human health and well-being (Díaz et al. 2019; IPBES 2022). However, it is worth mentioning that, if from one side urbanization poses several challenges, it also comes with several advantages. For example, in relation to mobility, dense urban areas can reduce the consumption of fossil fuel for private transport, and supports the soft-mobility; other relevant advantages can be related to the reduction of land use, artificial surface coverage, and energy consumption, as well as social equity (Chokhachian et al. 2019). Urban areas should therefore be planned and designed to face the challenges above mentioned by implementing a number of strategies to reduce the negative aspects of urbanization and optimize their efficiency. The strategies implemented ten, twenty years ago, or more, such as low-impact development (LID) and green infrastructures (GI), are now part of the broader definition of nature-based solutions (NbS). NbS are emerging as particularly effective in addressing the challenges mentioned above, improving human wellbeing and addressing a wide range of societal challenges (Cohen-Shacham et al. 2016).

1.2 Climate Change in Urban Areas

The issues related to climate change have been widely discussed in recent decades by the scientific community. Over the years, confidence in the role of anthropogenic activities, including land consumption, industrial activities, and related GHG emissions, has increased and was clearly demonstrated by the Intergovernmental Panel on Climate Change (IPCC) reports (IPCC 2018, 2021). Human-induced climate change has severe impacts on nature and people, beyond natural climate variability. Among the most relevant consequences of climate change, the increasing intensity of extreme events and related hazards poses a significant challenge. In this context, defining mitigation and adaptation strategies is relevant to understanding the overall content of the book. Mitigation strategies aim at reducing the intensity of the activities that drive climate change (i.e. climate change drivers), and, in particular, reduce greenhouse gas emissions. As defined by the IPCC, “Adaptation is the process of adjustment to actual or expected climate and its effects” (IPCC 2014). Adaptation strategies aim at reducing the risks related to climate change effects, such as extreme heat, rainfalls or drought, in order to improve urban resilience and reduce local vulnerability (Kundzewicz and Matczak 2012), and are essential for adjusting to the effects of climate change. The main objective of adaptation strategies is to reduce the vulnerability, although they might not be enough to face the irreversible impacts caused by increase of intense extreme events (Intergovernmental Panel on Climate Change (IPCC) 2023).

To effectively identify the most appropriate and effective adaptation and mitigation strategies, it is essential to provide a comprehensive overview of the drivers and driving forces of climate change, as well as the impacts of climate change on urban areas.

Climate change drivers encompass the physical processes that propel changes in climate, while driving forces refer to anthropogenic activities that underpin climate change. The Intergovernmental Panel on Climate Change's AR6 report (IPCC 2021) identifies greenhouse gas (GHG) emissions as the primary anthropogenic driver of climate change. These emissions lead to elevated concentrations of carbon dioxide (CO_2) and other pollutants, contributing to global temperature increases (Baker et al. 2018). Additionally, the Statistical Office of the European Union, in 2022, identified land use changes, economic growth, fuel combustion, and industrial processes as key driving forces behind climate change (Eurostat 2022).

The impacts of climate change in urban settings affect human health livelihoods and key infrastructure, especially in relation to the main effects: increased temperatures (Nie et al. 2024), altered precipitations (Mastrantonas et al. 2021), and GHG concentration and related air pollution (Sadatshojaie and Rahimpour 2020; Mosca et al. 2023). In particular, in relation to increasing temperature, cities features, such as impervious surfaces, reduced amount of vegetation, and building density, are the cause of intensification of heatwaves and local warming and runoff (Fig. 1.1, Kim and Brown 2021; Intergovernmental Panel On Climate Change (IPCC) 2023).

Given the complexity of cities and related environmental conditions, both adaptation and mitigation strategies are crucial (Marrone and Orsini 2018). According to the last IPCC report (Intergovernmental Panel on Climate Change (IPCC) 2023), the progress made in planning and implementing adaptation measures has resulted in numerous advantages. Effective adaptation strategies hold promise for mitigating climate-related risks and promoting sustainable development. In addition, the implementation of mitigation policies has contributed to the reduction of global energy and carbon intensity and greenhouse gas (GHG) emissions.

Adaptation is place- and context-specific (Field et al. 2014). A wide range of strategies and solutions can be implemented at different scales. In urban areas, adaptation solutions can be applied to buildings and open spaces to reduce one or more effects of climate change, starting at the building and district scale and extending to the urban scale (Fig. 1.2).

For example, green roofs and high vegetation, such as trees, on sun-exposed sides of buildings can help mitigate the effects of heatwaves and reduce the energy consumption for cooling indoor environments (Celik and Ogus Binatli 2018; Susca 2019). Urban gardens and groups of trees serve as both microclimate regulation

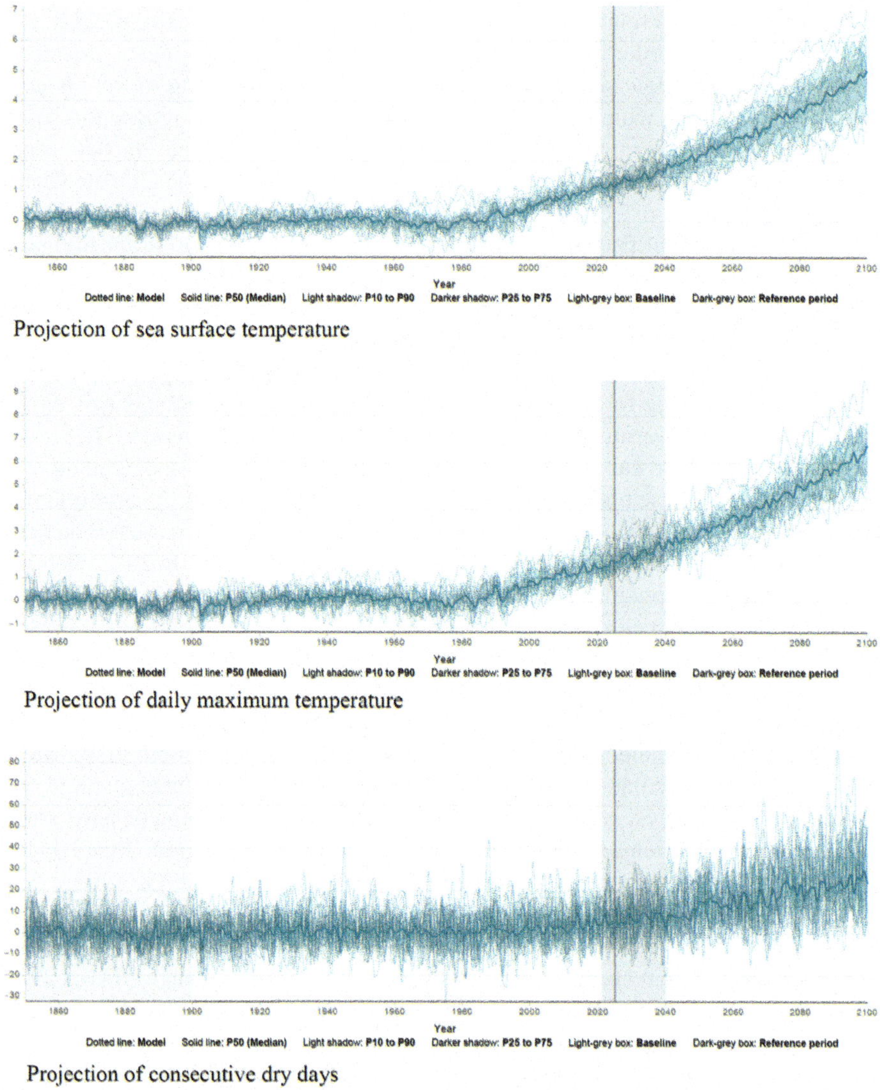

Fig. 1.1 Trend (1860–2020) and short-term projection (2021–2040) of sea surface temperature, maximum air temperature, and consecutive dry days, from the Coupled Model Intercomparison Project Phase 6 of the Copernicus Interactive Climate Atlas (O'Neill et al. 2016)

strategies and means to improve water management (Eisenberg et al. 2022; European Commission. Directorate General for Climate Action 2023).

Adaptation strategies hold significant potential for advancing the Sustainable Development Goals (SDGs) outlined in the United Nations' Agenda 2030. By

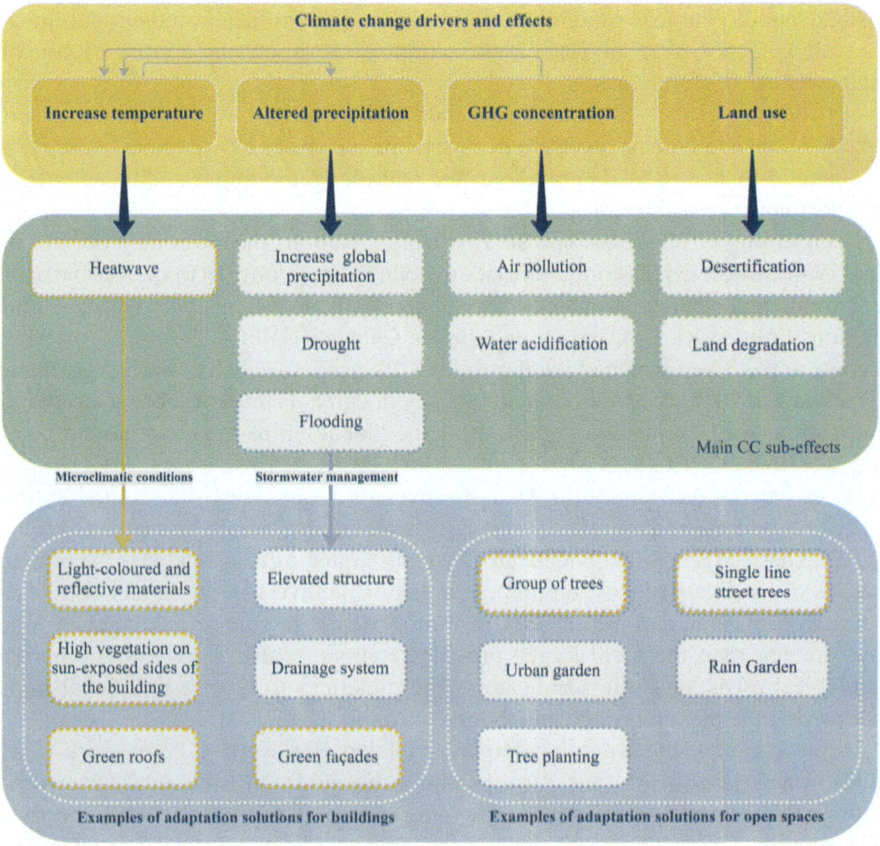

Fig. 1.2 Main climate change drivers and related consequences (Mosca et al. 2023), examples of adaptation strategies for the built environment (Eisenberg et al. 2022; European Commission. Directorate General for Climate Action 2023)

focusing on enhancing local resilience to environmental and climatic hazards, adaptation strategies aim to mitigate human and environmental risks (Meerow et al. 2016). These tailored solutions can directly and indirectly contribute to various Goals. For instance, implementing strategies to mitigate the impact of heatwaves through the integration of vegetation has a direct positive effect on Goal 3—Good Health and Well-being, as it reduces the occurrence of heat-related diseases (Lemus-Canovas et al. 2024).

Additionally, such initiatives align with Goal 11—Sustainable Cities and Communities, 13—Climate Action and Goal 15—Life on Land by fostering the introduction of plant species into urban environments, thereby supporting local biodiversity (Connop et al. 2016). In addition, depending on the scale of the intervention and on the selection of the type of strategy, these solutions can also indirectly support the achievement of other Goals such as the Goal 2—Zero hunger, by reducing the risks related to the loss of agricultural production due to heat, drought and floods, as well as Goal 6—Clean Water and Sanitation, by supporting a more sustainable management of stormwater.

In addition to the SDGs of the Agenda 2030, during the last decades, several international agreements on different environmental issues have been defined. These are mainly related to GHG emissions and anthropogenic impacts, climate change, and its impact on biodiversity.

Concerning GHG emissions, the first United Nations meetings date back to the late 1990s, where discussions took place to define a common plan to address the rising environmental challenges and climate change. With the establishment of the United Nations Framework Convention on Climate Change (UNFCCC), official meetings known as the 'Conference of the Parties' (COP) were created. The Paris Agreement, developed in 2015, constitutes one of the key documents in terms of climate change mitigation, with the main goal of limiting the global temperature increase to below 1.5 °C (UNFCCC 2015).

Focusing on biodiversity, the Aichi Biodiversity Targets, set by the United Nations in 2011 to address and mitigate biodiversity loss globally, should be mentioned (CBD 2012). These targets were divided into five main strategic goals, related to the primary causes of biodiversity loss, the improvement of biodiversity levels and ecosystem services provision, and the enhancement of strategies to support biodiversity. With this strategic plan, the Aichi Targets provide a comprehensive overview of the current ecological crisis, and offers indicators and guidelines to reverse the global trend, addressing the most relevant challenges.

More recently, the Kunming-Montreal Global Biodiversity Framework (CBD 2022) was established by the Parties to the Convention on Biological Diversity to replace the previous Aichi Targets, providing more detailed and concrete objectives. For example, key focus areas include endangered species, the restoration of degraded areas across various ecosystems, and minimizing the impact of pollutants and climate change on ecosystems.

Finally, COP15 in 2024, also known as the UN Biodiversity Conference, highlighted the ongoing global environmental crisis and emphasized the crucial role of nature-based solutions in improving the quality of both urban and natural ecosystems (UNFCCC 2024).

The role of nature-based solutions, given their versatility in terms of scale, performance, and implementability, is therefore becoming increasingly central to achieving the aforementioned international objectives and targets, as well as supporting local strategic development plans.

1.3 Nature-Based Cities

The rising urban complexity can be effectively managed through computational modelling (Peters and Peters 2018). In Nature-based Cities we focus on methodologies and tools to design nature-based solutions (NbS) for climate change adaptation, particularly for microclimate regulation and stormwater retention, addressing site-specific challenges.

Nature-based solutions are defined as solutions inspired by nature (European Commission 2016), and they can be either solutions that involve the direct use of natural elements such as plants or artificial solutions that, however, are inspired by natural processes of ecosystems.

Nature-based solutions are widely recognized as one of the most significant Climate Change (CC) adaptation strategies (Gaffin et al. 2012; Kabisch et al. 2016; Derkzen et al. 2017) due to their versatility and ability to manage various CC effects such as temperature increases and alterations in precipitation patterns.

Concerning the improvement of the microclimate, two primary categories of NbS can positively impact microclimatic conditions. The first category involves the use of surface materials with high albedo levels, which is defined as the material's capacity to reflect a fraction of incident light (Coakley 2003). This approach can effectively reduce local surface temperatures and mitigate the urban heat island (UHI) effect (Akbari et al. 2015). The second category focuses on the introduction of vegetation, which can enhance microclimatic conditions through several mechanisms. Vegetation provides shade, influencing local air temperature, and facilitates evaporation and transpiration processes, which in turn affect relative humidity and mean radiant temperature (Abdulateef and Al-Alwan 2021), as deeply described in Chaps. 2 and 3.

Simultaneously, one of the most pressing urban challenges revolves around mitigating the risks of flooding. With Climate Change, the frequency of extreme rainfall events is projected to increase, exacerbating the inefficiency of local sewage systems in handling varying stormwater volumes (Qin 2020). Nature-based solutions offer valuable support in reducing flooding risks and managing stormwater. They achieve this by locally mitigating water runoff, thereby decreasing the volume of stormwater that overburdens local sewage systems and infrastructure. The most influencing factors and relevant parameters are described in Chaps. 2 and 4.

Solutions such as raingardens, green roofs, permeable pavements, and tree planting demonstrate effectiveness at the local level (Eisenberg et al. 2022; European Commission. Directorate General for Climate Action. 2023). However, the design of these solutions should be tailored to the specific microclimatic conditions, rainfall patterns, and urban morphology and configuration of each site.

The performances of NbS are site-specific and depend on several factors (Directorate-General for Research and Innovation (European Commission) 2021), which in this book are systematically discussed and addressed in a computational design workflow (Chap. 5), to present a methodology that can be used by designers with different levels of complexity, discussing its applicability and limitations according to selected Key Performance Indicators and related thresholds.

More specifically, the chapter on **Key Performances of Nature-based Solutions** (Chap. 2), outlines the main performances of nature-based solutions concerning the ecological and environmental issues of cities and the impacts of climate change. It includes aspects such as microclimate regulation, air quality improvement, water management, and highlights the natural processes of NbS in supporting the improvement of urban conditions in relation to these environmental challenges. Relevant Key Performance Indicators (KPIs) are listed accordingly. For each KPI, the most influential parameters/characteristics of NbS are provided, as per the existing literature. The focus is primarily on microclimate and human comfort conditions improvement and stormwater retention.

The chapter **Nature-based Solutions for Microclimate Regulation and Human Comfort** (Chap. 3), explains in more detail the performance of NbS in mitigating the urban heat island effect and improving thermal comfort conditions for humans. In addition, the main characteristics and processes on which these performances depend are explored. Then, this chapter introduces a wide range of currently available computational approaches and tools to model and simulate the performances of NbS for regulating microclimate and improving outdoor thermal comfort in relation to the aforementioned climate change impacts and environmental issues. An overview of existing tools and an in-depth analysis of the input parameters needed for modelling and simulating is described, and the expected outcomes are illustrated. Among the described tools, the most relevant and effective for simulating NbS in the urban environment are selected, and a series of applications on general case studies is showcased.

The chapter **Nature-based Solutions for Stormwater Retention** (Chap. 4), illustrates the complexity of the water cycle in cities and the anthropic activities and processes driving the alteration of precipitations. The main risks and challenges related to extreme events caused by altered precipitation are described, and the role of NbS in mitigating related risks are highlighted. Then, this chapter presents an overview of existing computational approaches and tools to model and simulate the performances of NbS for stormwater retention, addressing the mentioned climate change impacts and environmental issues. The chapter provides an overview of existing tools and offers a detailed description of the input parameters required for analysis and simulation. Among the tools listed, two approaches are selected according to design and technical requirements, and their application on typical urban canyons (consistent with the ones considered in Chap. 3), is showcased and discussed.

In the chapter **Performance-Driven Design Approach: a case study** (Chap. 5), a comprehensive workflow for a coupled performance-driven design approach is presented for readers, offering a pathway to define design strategies based on site-specific Key Performance Indicators (KPIs). A case study on a real site in the Mediterranean area is included to better explain the (computational and theoretical) approach and present research results. By applying the implemented workflow step by step to three different design options for the same site and design requirements, each with

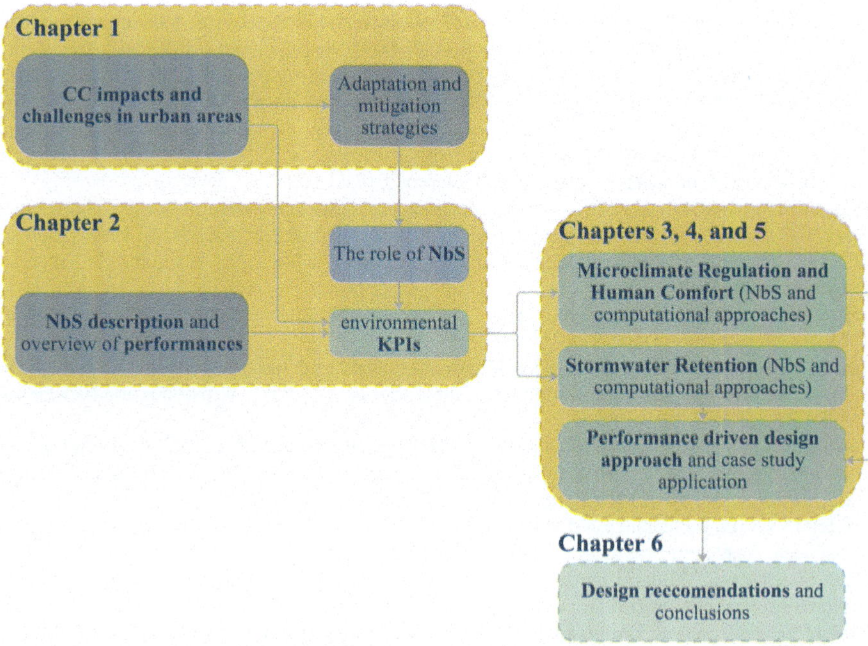

Fig. 1.3 Structure and main content of the book

several NbS options, the main information, data, and inputs required are illustrated, and the different outputs provided are described. Finally, a detailed analysis of the results for the three configurations is provided, in relation to the selected KPIs.

In the final chapter, **Design recommendations for Nature-based Cities** (Chap. 6), a summary of the potential role of NbS in mitigating environmental and ecological crisis and managing climate-related risk is provided; in addition, starting from the results provided and discussed in the previous chapters, design recommendations for microclimate regulation and for stormwater management improvement through the integration of NbS are provided, and general criteria for design optimization are derived, in relation to different urban conditions.

Figure 1.3 shows the structure of Nature-based Cities and the relation between the topics addressed in the different chapters.

References

Abdulateef MF, Al-Alwan HA (2021) The effectiveness of urban green infrastructure in reducing surface urban heat island. Ain Shams Eng J

Akbari H, Cartalis C, Kolokotsa D et al (2015) Local climate change and urban heat island mitigation techniques—the state of the art. J Civil Eng Manag 22:1–16. https://doi.org/10.3846/13923730.2015.1111934

Aronson MF, Lepczyk CA, Evans KL et al (2017) Biodiversity in the city: key challenges for urban green space management. Front Ecol Environ 15:189–196. https://doi.org/10.1002/fee.1480

Baker HS, Millar RJ, Karoly DJ et al (2018) Higher CO_2 concentrations increase extreme event risk in a 1.5 °C world. Nature Clim Change 8:604–608. https://doi.org/10.1038/s41558-018-0190-1

Beheshtian A, Donaghy KP, Gao HO et al (2018) Impacts and implications of climatic extremes for resilience planning of transportation energy: a case study of New York city. J Clean Prod 174:1299–1313. https://doi.org/10.1016/j.jclepro.2017.11.039

Benedict MA, McMahon ET (2001) Green infrastructure: smart conservation for the 21st century

CBD (2012) Aichi biodiversity targets. https://www.cbd.int/sp/targets. Accessed 18 Mar 2025

CBD (2022) Kunming-Montreal global biodiversity framework. https://www.cbd.int/gbf. Accessed 18 Mar 2025

Celik S, Ogus Binatli A (2018) Energy savings and economic impact of green roofs: a pilot study. Emerg Mark Financ Trade 54:1778–1792

Chen K, Boomsma J, Holmes HA (2023) A multiscale analysis of heatwaves and urban heat islands in the western U.S. during the summer of 2021. Sci Rep 13:9570. https://doi.org/10.1038/s41598-023-35621-7

Chokhachian A, Perini K, Giulini S, Auer T (2019) Urban performance and density: generative study on interdependencies of urban form and environmental measures. Sustain Cities Soc 101952. https://doi.org/10.1016/j.scs.2019.101952

Coakley JA (2003) Reflectance and albedo, surface. In: Holton JR (ed) Encyclopedia of atmospheric sciences. Academic Press, Oxford, pp 1914–1923

Cohen-Shacham E, Walters G, Janzen C, Maginnis S (eds) (2016) Nature-based solutions to address global societal challenges. IUCN International Union for Conservation of Nature

Connop S, Vandergert P, Eisenberg B et al (2016) Renaturing cities using a regionally-focused biodiversity-led multifunctional benefits approach to urban green infrastructure. Environ Sci Policy 62:99–111. https://doi.org/10.1016/j.envsci.2016.01.013

Derkzen ML, van Teeffelen AJA, Verburg PH (2017) Green infrastructure for urban climate adaptation: how do residents' views on climate impacts and green infrastructure shape adaptation preferences? Landsc Urban Plan 157:106–130. https://doi.org/10.1016/j.landurbplan.2016.05.027

Díaz S, Settele J, Brondízio ES et al (2019) Pervasive human-driven decline of life on Earth points to the need for transformative change. Science 366:eaax3100. https://doi.org/10.1126/science.aax3100

Directorate-General for Research and Innovation (European Commission) (2021) Evaluating the impact of nature-based solutions: a handbook for practitioners. Publications Office of the European Union, LU

Du WC, Xia XH (2018) How does urbanization affect GHG emissions? A cross-country panel threshold data analysis. Appl Energy 229:872–883. https://doi.org/10.1016/j.apenergy.2018.08.050

Eisenberg B, Chiesa C, Fischer LK et al (2022) Nature-based solutions technical handbook factsheets

European Commission (2016) Nature-based solutions. https://ec.europa.eu/research/environment/index.cfm?pg=nbs. Accessed 3 Oct 2016

European Environment Agency (2022) Air quality in Europe 2022

European Commission. Directorate General for Climate Action. (2023) EU-level technical guidance on adapting buildings to climate change: best practice guidance. Publications Office, LU

Eurostat (2022) Climate change—driving forces

Field CB, Barros VR, Mastrandrea MD et al (2014) Climate change 2014: summary for policymakers. Climate change 2014: impacts, adaptation, and vulnerability part A: global and sectoral aspects. In: Contribution of working group II to the fifth assessment report of the intergovernmental panel on climate change, pp 1–32

Gaffin SR, Rosenzweig C, Kong AYY (2012) Adapting to climate change through urban green infrastructure. Nature Clim Change 2:704–704. https://doi.org/10.1038/nclimate1685

Grimm NB, Grove JG, Pickett STA, Redman CL (2000) Integrated approaches to long-term studies of urban ecological systems: urban ecological systems present multiple challenges to ecologists—pervasive human impact and extreme heterogeneity of cities, and the need to integrate social and ecological approaches, concepts, and theory. Bioscience 50:571–584. https://doi.org/10.1641/0006-3568(2000)050[0571:IATLTO]2.0.CO;2

Groffman PM, Avolio M, Cavender-Bares J et al (2017) Ecological homogenization of residential macrosystems. Nature Ecol Evol 1:1–3. https://doi.org/10.1038/s41559-017-0191

Intergovernmental Panel On Climate Change (IPCC) (2023) Climate change 2022—impacts, adaptation and vulnerability: working group II contribution to the sixth assessment report of the intergovernmental panel on climate change, 1st edn. Cambridge University Press

IPBES (2022) Global assessment report on biodiversity and ecosystem services of the Intergovernmental Science-Policy Platform on Biodiversity and Ecosystem Services. Zenodo. URL https://zenodo.org/record/5657041

IPCC (ed) (2015) Climate change 2014: synthesis report. Intergovernmental Panel on Climate Change, Geneva, Switzerland

IPCC (2018) Global Warming of 1.5°C. An IPCC Special Report on the impacts of global warming of 1.5°C above pre-industrial levels and related global greenhouse gas emission pathways, in the context of strengthening the global response to the threat of climate change, sustainable development, and efforts to eradicate poverty (Masson-Delmotte V, Zhai P, Pörtner H-O, Roberts D, Skea J, Shukla PR, Pirani A, Moufouma-Okia W, Péan C, Pidcock R, Connors S, Matthews JBR, Chen Y, Zhou X, Gomis MI, Lonnoy E, Maycock T, Tignor M, Waterfield T (eds)). Cambridge University Press, Cambridge, UK and New York, NY, USA, 616 pp

IPCC (2021) Climate change 2021. The Physical Science Basis. https://www.ipcc.ch/report/ar6/wg1/downloads/report/IPCC_AR6_WGI_Full_Report.pdf. Accessed 27 Oct 2021

Jamali B, Bach PM, Deletic A (2020) Rainwater harvesting for urban flood management—an integrated modelling framework. Water Res 171:115372. https://doi.org/10.1016/j.watres.2019.115372

Kabisch N, Frantzeskaki N, Pauleit S et al (2016) Nature-based solutions to climate change mitigation and adaptation in urban areas: perspectives on indicators, knowledge gaps, barriers, and opportunities for action. Ecol Soc 21

Kim SW, Brown RD (2021) Urban heat island (UHI) variations within a city boundary: a systematic literature review. Renew Sustain Energy Rev 148:111256. https://doi.org/10.1016/j.rser.2021.111256

Kundzewicz ZW, Matczak P (2012) Natural risks: mitigation and adaptation. Ecohydrol Hydrobiol 12:3–8. https://doi.org/10.2478/v10104-012-0005-3

Lederer J, Gassner A, Fellner J et al (2021) Raw materials consumption and demolition waste generation of the urban building sector 2016–2050: A scenario-based material flow analysis of Vienna. J Clean Prod 288:125566. https://doi.org/10.1016/j.jclepro.2020.125566

Lemus-Canovas M, Insua-Costa D, Trigo RM, Miralles DG (2024) Record-shattering 2023 Spring heatwave in western Mediterranean amplified by long-term drought. NPJ Clim Atmos Sci 7:25. https://doi.org/10.1038/s41612-024-00569-6

Maes J, Zulian G, Thijssen M et al (2016) Mapping and assessment of ecosystems and their services—urban ecosystems 4th report

Marrone P, Orsini F (2018) Resilience and open urban environments. Comparing adaptation and mitigation measures. J Technol Arch Environ 348–357. https://doi.org/10.13128/TECHNE-22099

Mastrantonas N, Herrera-Lormendez P, Magnusson L et al (2021) Extreme precipitation events in the Mediterranean: Spatiotemporal characteristics and connection to large-scale atmospheric flow patterns. Int J Climatol 41:2710–2728. https://doi.org/10.1002/joc.6985

McDonald RI, Mansur AV, Ascensão F et al (2020) Research gaps in knowledge of the impact of urban growth on biodiversity. Nat Sustain 3:16–24. https://doi.org/10.1038/s41893-019-0436-6

Meerow S, Newell JP, Stults M (2016) Defining urban resilience: a review. Landsc Urban Plan 147:38–49. https://doi.org/10.1016/j.landurbplan.2015.11.011

Mosca F, Canepa M, Perini K (2023) Strategies for adaptation to and mitigation of climate change: key performance indicators to assess nature-based solutions performances. Urban Clim 49:101580. https://doi.org/10.1016/j.uclim.2023.101580
Nelson EJ, Kareiva P, Ruckelshaus M et al (2013) Climate change's impact on key ecosystem services and the human well-being they support in the US. Front Ecol Environ 11:483–893. https://doi.org/10.1890/120312
Nie G, Yang J, Zhang Y et al (2024) Duration of exposure to compound daytime-nighttime high temperatures and changes in population exposure in China under global warming. Hum Soc Sci Commun 11:1065. https://doi.org/10.1057/s41599-024-03586-1
O'Neill BC, Tebaldi C, Van Vuuren DP et al (2016) The scenario model intercomparison project (ScenarioMIP) for CMIP6. Geosci Model Dev 9:3461–3482. https://doi.org/10.5194/gmd-9-3461-2016
Peters B, Peters T (2018) Computing the environment: digital design tools for simulation and visualisation of sustainable architecture. Wiley, Chichester, West Sussex
Qin Y (2020) Urban flooding mitigation techniques: a systematic review and future studies. Water 12:3579. https://doi.org/10.3390/w12123579
Rashid N, Alam JAMM, Chowdhury MdA, Islam SLU (2022) Impact of landuse change and urbanization on urban heat island effect in Narayanganj city, Bangladesh: a remote sensing-based estimation. Environ Challen 8:100571. https://doi.org/10.1016/j.envc.2022.100571
Sadatshojaie A, Rahimpour MR (2020) CO_2 emission and air pollution (volatile organic compounds, etc.)–related problems causing climate change. In: Current trends and future developments on (bio-) membranes. Elsevier, pp 1–30
Susca T (2019) Green roofs to reduce building energy use? A review on key structural factors of green roofs and their effects on urban climate. Build Environ 162:106273
Taha H (2008) Meso-urban meteorological and photochemical modeling of heat island mitigation. Atmos Environ 42:8795–8809. https://doi.org/10.1016/j.atmosenv.2008.06.036
UNFCCC (2015) The Paris agreement. https://unfccc.int/process-and-meetings/the-paris-agreement. Accessed 18 Mar 2025
UNFCCC (2024) Nature for climate. https://unfccc.int/news/nature-for-climate. Accessed 18 Mar 2025
United Nations (2012) Managing water report under uncertainty and risk. The United Nations World Water Development Report 4. United Nations Educational, USA
United Nations, Department of Economic and Social Affairs, Population Division (2018) World urbanization prospects: the 2018 revision
Wu J (2008) Making the case for landscape ecology: an effective approach to urban sustainability. Landsc J 27:41–50. https://doi.org/10.3368/lj.27.1.41

Chapter 2
Key Performances of Nature-Based Solutions

Abstract This chapter outlines the main benefits of nature-based solutions (NbS) in addressing the ecological and environmental challenges of cities, as well as the impacts of climate change and related risks to humans and ecosystems. It covers aspects such as microclimate regulation, air quality improvement, and water management. The key ecosystem services provided by NbS are listed in relation to the relevant Sustainable Development Goals. The contribution of nature-based solutions to climate change adaptation, risk management, and prevention is broadly discussed. Relevant Key Performance Indicators (KPIs) are presented accordingly. For each KPI, the most influential parameters and characteristics of NbS are described, based on existing literature. The focus is primarily on human comfort and stormwater retention.

Keywords Nature-based solutions · Environmental performance · KPI · Microclimate · Thermal comfort · Water management · Benefits · Ecosystem services

2.1 Nature-Based Solutions for CC Adaptation

Nature-based solutions (NbS) are solutions inspired and/or supported by nature. NbS are defined as "interventions which use nature and the natural functions of healthy ecosystems to tackle some of the most pressing challenges of our time. These types of solutions help to protect the environment but also provide numerous economic and social benefits" (IUCN, International Union for Conservation of Nature 2020; UNEA 2022). Nature-based Solutions provide a wide range of ecosystem services supporting climate change adaptation, contributing to risk management and prevention. Ecosystem services are benefits provided by nature playing a key role in improving the ecological and environmental conditions of cities (Costanza and Mageau 1999; Maes et al. 2016b; IUCN 2016).

Ecosystem services are classified into four categories (Millennium Ecosystem Assessment (Program) 2005; Maes et al. 2016a, b): (1) provisioning services;

K. Perini and F. Mosca, *Nature-Based Cities*, Digital Innovations in Architecture, Engineering and Construction, https://doi.org/10.1007/978-3-031-94612-7_2

(2) regulating services; (3) support services, (4) cultural services. Provisioning services include the supply of food and other raw materials, such as wood for the building sector. Supporting services sustain essential ecosystem processes, including providing habitats for plants and wildlife, as well as preserving genetic and biological diversity. Cultural services are also highly related to the design and planning of cities, as they focus on social relations, health, and wellbeing. Finally, regulating services, which are the focus of this chapter, include climate regulation and water-related services, and are linked to the effects that nature-based solutions can have on climate change impacts such as heatwaves and flooding (Mosca et al. 2023). Such impacts constitute a major environmental risk, as highlighted by the Global Risks Report (World Economic Forum 2024). Indeed, according to the Report, the most significant environmental risks in the short (2 years) and long (10 years) term include extreme weather events, critical change to Earth systems, pollution, biodiversity loss and ecosystem collapse. Also, other ecosystem services can contribute to improving the management and prevention of environmental risks. Nature-based solutions can improve air quality and water quality, managing or preventing pollution risks. Vegetation provides oxygen in exchange for carbon dioxide and ecosystem service for removing air pollutants (Chen et al. 2023; Haq et al. 2024), such as particulate matters (PM10 and PM2.5). The construction of greenways and urban parks in cities can also reduce the use of private transportation, promoting active mobility (Lemieux et al. 2023). Water cycle renaturalisation, thanks to canopy interception, flow control, flood reduction, soil infiltration and storage, and water quality improvement, strongly contributes in managing and preventing risks such as pollution, extreme weather events, and critical change to earth systems (Perini and Sabbion 2016). Finally, the risks of biodiversity loss and ecosystem collapse can be addressed through the planning of nature-based solutions at different scales, as they can promote biodiversity conservation and enhancement, create habitats and corridors, and foster species diversity (McPhearson et al. 2015; Perini and Sabbion 2016; Aronson et al. 2017).

In addition, Wood et al. (2018) highlight that regulating services play a crucial role in advancing the UN Sustainable Development Goals (SDGs, United Nations 2015). For instance, regulating local climate and air quality strongly supports SDG 3 Good Health and Well-being, particularly target 3.9, which focuses on environmental health aimed at reducing the number of deaths and illnesses from hazardous chemicals, as well as air, water, and soil pollution and contamination contribute to SDG 11 Sustainable Cities and Communities and 13 Climate Action. Similarly, the moderation of extreme events contributes to SDGs 2 Zero Hunger, 6 Clean Water and Sanitation, 13 Climate Action, and 14 Life on Land.

The contribution of nature-based solutions to achieving SDG 11 (Sustainable Cities and Communities) is mainly related to the following targets:

- 11.3: Inclusive and Sustainable Urbanization—this target aims to enhance inclusive and sustainable urbanization and the capacity for participatory, integrated, and sustainable human settlement planning and management in all countries.

- 11.4: Protect the World's Cultural and Natural Heritage—Strengthen efforts to protect and safeguard the world's cultural and natural heritage.
- 11.5: Reduce the Adverse Effects of Natural Disasters—Reduce the number of deaths and the number of people affected by disasters, and substantially decrease direct economic losses relative to global gross domestic product caused by disasters, including water-related disasters, with a focus on protecting the poor and people in vulnerable situations.
- 11.6: Reduce the Environmental Impact of Cities—This target includes reducing the environmental impact of cities, with particular attention to air quality and municipal and other waste management.
- 11.7: Provide Access to Safe and Inclusive Green and Public Spaces—Ensure access to safe and inclusive green and public spaces for all.

Focusing on SDG 13 Climate Action, the most relevant target related to the integration of nature-based solutions in cities is 13.1: Strengthen Resilience and Adaptive Capacity to Climate-Related Disasters, as discussed above when addressing risk management and prevention. Additionally, the other targets 13.2: Integrate Climate Change Measures into Policies and Planning and 13.3: Build Knowledge and Capacity to Meet Climate Change can be linked to the cultural ecosystem services provided by nature-based solutions.

Nature-based solutions can be divided into: built structure, water media and land media (Babí Almenar et al. 2021) (Fig. 2.1). These categories allow for outlining the various key performances, in terms of ecosystem services provided, which can be achieved through site-specific and context-sensitive urban planning, urban design, and building design across different scales.

The NbS built structure category includes green roofs and vertical greening systems (Fig. 2.2). Such systems allow integrating vegetation in the building envelope, thanks to a range of layers. Green roofs can be intensive and extensive, providing slightly different benefits, for example, the first type is accessible, the second not (Pérez and Coma 2018). Vertical greening systems can be classified according to their growing method in green façades, based on the use of climbing plants, or living walls. The latter provides higher performances, such as energy savings, but require higher installation and maintenance costs (Coma et al. 2017; Fernández-Cañero et al. 2018).

Examples of land-based nature-based solutions (NbS) include tree groups, meadows, and orchards, providing a wide range of performances depending on the amount and type of plant (Fig. 2.3) (Hirata et al. 2024). Finally, water media NbS include systems specifically designed and built to improve water management, such as rain gardens, swales (Fig. 2.4) and permeable pavements (Ballard et al. 2007; VirginiaTech 2013). Water-based systems can be vegetated, as in the case of rain gardens, or made of artificial materials to filter and infiltrate stormwater, as in the case of permeable pavements.

Figure 2.5 shows the main ecosystem services provided by different types of NbS for each category, in relation to the CC drivers and effects (outlined in Chap. 1) (Eisenberg et al. 2022; European Commission. Directorate General for Climate Action

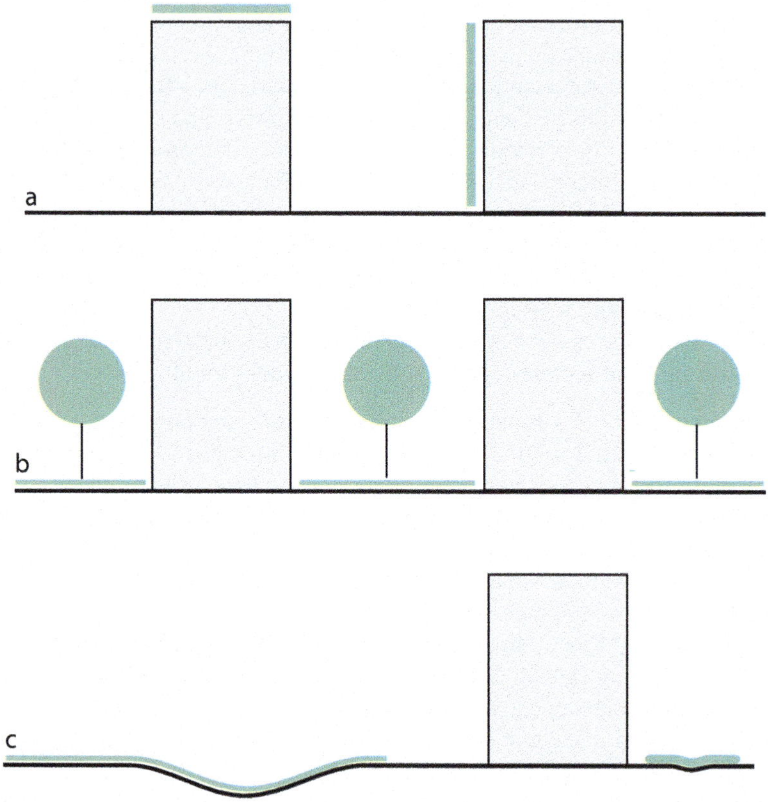

Fig. 2.1 Nature-based solutions, classified according to Babí Almenar et al. 2021: **a** built structure: green roof and vertical greening; **b** land media: tree groups, meadows; **c** water media: swale and rain garden

2023). The performances in terms of regulating services of built structures are mainly related to local climate, air quality regulation (Alexandri and Jones 2008; Castleton et al. 2010; Baik et al. 2012) and carbon sequestration and storage (Agra et al. 2017). Land media provide all the main regulating services, with performances depending on the amount of biomass. For example, meadows and tree groups will contribute differently to local climate regulation (Mosca et al. 2021). More specifically, the overall performances of tree groups depend on foliage shape and dimensions, height of trunk, leaf area density (LAD), seasonal cycle, particularly in the case of microclimatic regulation (Scudo and Ochoa de la Torre 2003; Grosso 2012; Perini et al. 2018a).

Finally, vegetated water media NbS provide a range of ecosystem services, such as local and air quality regulation, erosion prevention, carbon sequestration and storage, besides moderation of extreme events, while others non-vegetated, such as permeable pavements, contribute to moderation of extreme events and wastewater only. Their performance depends on the layers and on the existing and site specific conditions

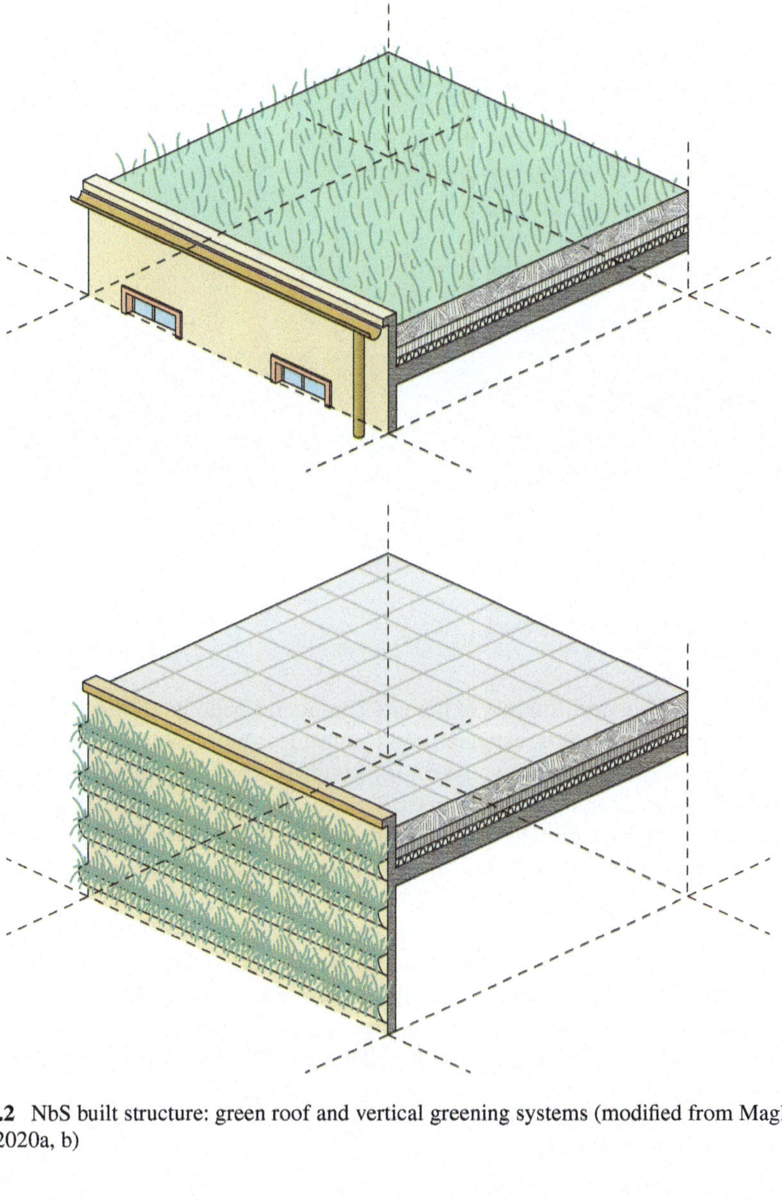

Fig. 2.2 NbS built structure: green roof and vertical greening systems (modified from Magliocco et al. 2020a, b)

which determine the possibility of infiltration of rain water (Magliocco et al. 2020a, b).

Nature-based solutions integrated into urban areas can have different characteristics and specific features, depending on the climatic region. For example, a green wall can be simply made of climbing plants covering a building's façade (Fig. 2.6). There are numerous examples dating back to the nineteenth century of such systems in European and North American cities, where woody climbers were commonly used

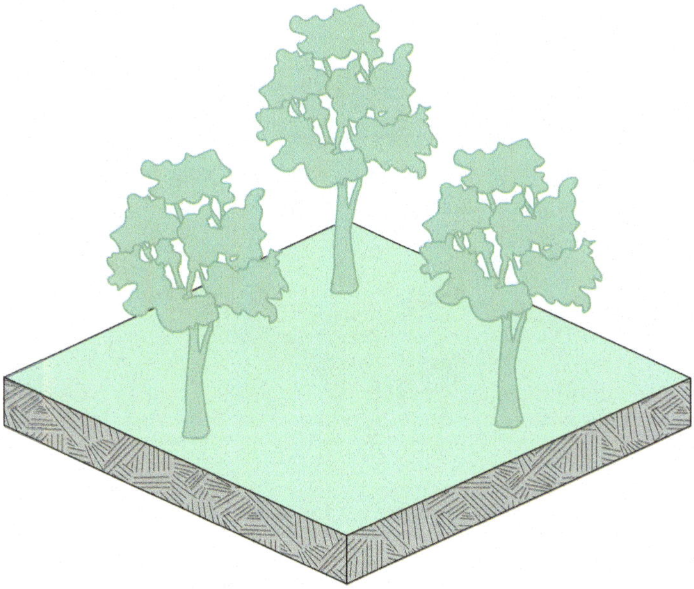

Fig. 2.3 NbS land media: tree groups

to cover simple façades. Interest in the environmental benefits of these systems can be traced back to the 1980s, especially in Central Europe (Köhler 2008).

Figure 2.7 shows an intensive green roof built in Porto (Portugal) on the former Mercado do Anjo, hosting 50 olive trees. This regeneration process, which began in 2007 and was completed in 2013, demonstrates how a green roof can become a new space for the community, also creating a pedestrian architectural path that connects two important areas of the city (Urban Nature Atlas 2021). This example illustrates that, although classifying NbS types can be very useful for planning, design, and research purposes, case studies can sometimes transcend such typologies.

In the case of the NbS presented for the city of Munich (Fig. 2.8), it is interesting to note that the land-based NbS (meadow and trees) are connected to a semi-intensive green roof, allowing people to walk on it. The combination of these solutions have been implemented for the subway station of the newest and biggest Campus of the Technical University of Munich, in Garching. The implementation of these solutions highlights the role of NbS in providing green areas for recreation not only in outdoor spaces such as parks and avenues, but also nearby infrastructure systems, such as public transport stations.

The last case study presented is a rain garden, a water media NbS, built in Campomorone, near the city center of Genoa (Italy). This NbS is a pilot project built as part of the "Proterina3 Evolution" strategic project, funded by the Interreg Maritime Italy-France program. This rain garden was designed to be monitored through the use of a weather station, a drainage system with a level sensor, and an electric valve. For this purpose, a tank is placed to collect rainwater from a terrace covering a car garage

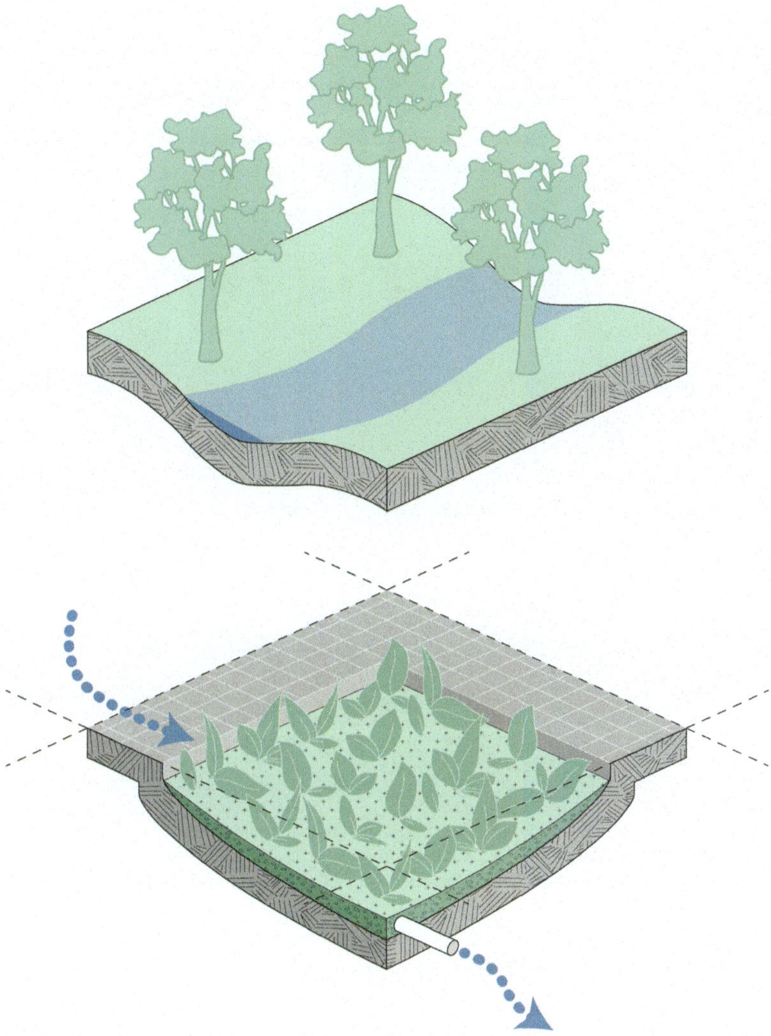

Fig. 2.4 NbS water media: swale and rain garden (modified from Magliocco et al. 2020a, b)

located nearby, as well as manholes connecting the various elements. The rain garden surface is 16% of collective surface. Defining this area is possible by applying calculation methods defined in several guidelines, starting from the precipitation regime of the site. Additionally, it was designed as a non-infiltrating rain garden, with the lower surface of the excavation impermeabilized before being filled with soil. This design choice is related to the proximity of a school building (to avoid any structural problem due to soil erosion) and the need to monitor, for research purposes, the inlet and outlet water to quantify the stormwater retention of the system (Fig. 2.9, Magliocco et al. 2020a, b).

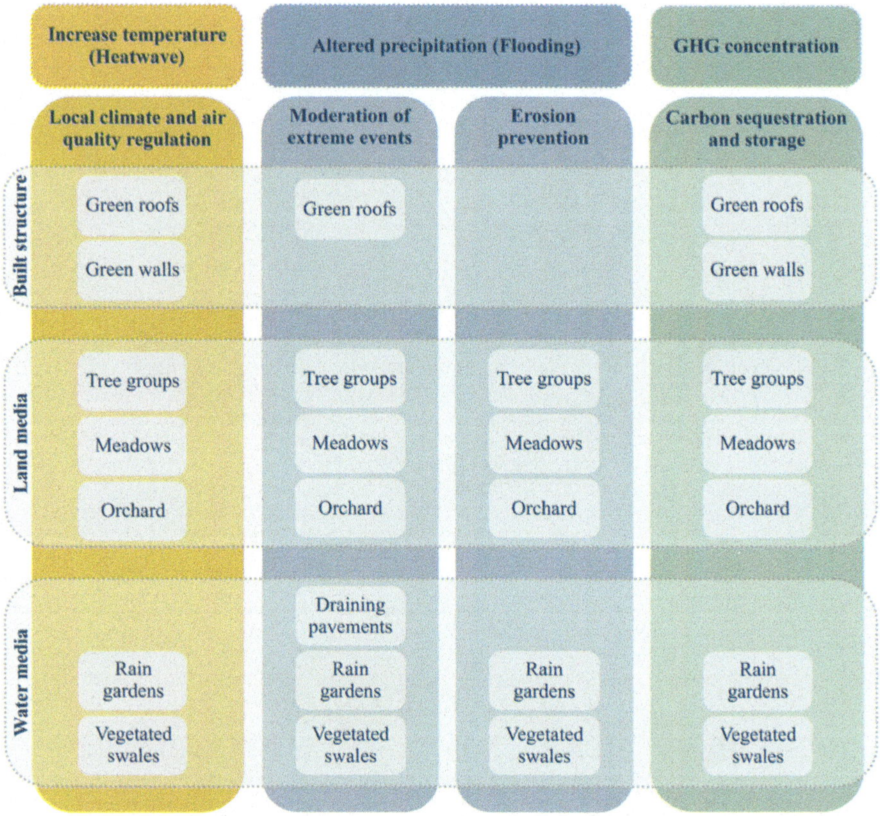

Fig. 2.5 Regulating services for CC adaptation provided by Nature-based solutions, classified according to Babí Almenar et al. (2021), in relation to CC drivers and effects (Mosca et al. 2023)

2.2 Microclimate Regulation

The impact of climate change, particularly the increase in temperature and its sub-effect, heatwaves (Masson et al. 2020), significantly affects urban areas also due to the urban heat island (UHI) phenomenon. This phenomenon arises from the prevalence of artificial surfaces with low albedo, anthropogenic heat emissions, reduced evaporation, and lower wind speeds caused by high building density (Oke 1982; Taha 1997; Grimmond 2007). In this context, elevated temperatures in cities can severely impact human health (Morabito et al. 2015; Thompson et al. 2023; Li 2024).

Microclimate regulation falls under the ecosystem service of local climate and air quality regulation, particularly at smaller scales such as neighbourhoods and individual buildings. Nature-based solutions play a key role in microclimate regulation through mechanisms such as shading, evapotranspiration, and other processes (Manickathan et al. 2018; Bakhshoodeh et al. 2022).

Fig. 2.6 Vertical greening system based on climbing plants in Copenhagen, Denmark

The most relevant effects, mechanisms/methods and related NbS influencing parameters are summarized in Fig. 2.10. Shading allows intercepting solar radiation, thus limiting the heating of any type of surfaces (e.g. asphalt road, characterised by low albedo values (Taha 1997). The shading capacity depends on the density of the foliage and its shape and dimensions. This latter needs to be considered, in the design

Fig. 2.7 Intensive green roof system with planted trees in Porto, Portugal

process, depending on the latitude of the site (Grosso 2012). Evapotranspiration by plants reduces air temperature, depending on the climatic conditions (e.g. mean daily solar radiation, mean daily temperature, wind speed (Zotarelli et al. 2024) and plant species' foliage, seasonal cycle and daily transpiration (Scudo and Ochoa de la Torre 2003). Finally, wind speed can be modified by vegetation to improve microclimatic conditions (Perini et al. 2018a).

Depending on the type of NbS, the most influential parameters can play different roles (Fig. 2.11). For built structures, intensive green roofs can be designed to maximize foliage density, shape, dimensions, and daily transpiration. The seasonal cycle is primarily associated with intensive green roofs that host trees. For extensive green roofs, these parameters are less relevant, and their contribution to microclimate regulation is mainly linked to the reduction of mean radiant temperature (Coma et al. 2018; Pérez and Coma 2018). Living wall systems can have a dense foliage, contributing

Fig. 2.8 Semi-intensive green roof system and land-media NbS in Munich, Germany

to microclimate regulation, while green façades can integrated deciduous species to allow direct solar radiation reaching a building envelope during winter seasons (Coma et al. 2017).

In the case of NbS land media (Fig. 2.12), orchards perform similarly to extensive green roofs, mainly reducing mean radiant temperature. However, tree group designs should consider various parameters to optimize microclimate regulation

Fig. 2.9 Rain Garden, pilot project of Proterina 3 Evolution, Genoa, Italy (credit: Gianluca Porcile)

performance. Foliage density, shape, and dimensions can vary significantly, and high performance can be achieved through careful design of tree groups. For example, a line of trees can effectively shade a street or building, leading to increased thermal comfort (Perini et al. 2018a). Additionally, the seasonal cycle can play a crucial role in ensuring thermal comfort during the winter season (Scudo and Ochoa de la Torre 2003).

Finally, water-based NbS can also contribute to microclimate regulation (Fig. 2.13). Rain gardens can be designed with plant species that have high foliage density and high daily transpiration rate. However, it is important to note that these systems cannot integrate trees due to the layers involved at the ground level, which are necessary to ensure filtration and infiltration performance (Godwin et al. 2011; Magliocco et al. 2020a, b).

2.3 Water Management

In the field of climate change adaptation, water management is of key importance, especially in cities. Among the Nature-based solutions analyzed in this chapter, green walls were not considered, as their vertical geometry limits their contribution (unless they are designed with a rainwater-based irrigation system).

Figure 2.14 shows the main treatment methods for the different NbS and highlights the mechanisms and related influencing parameters. The treatment methods are identified according to treatment train approach, which describes the management of both

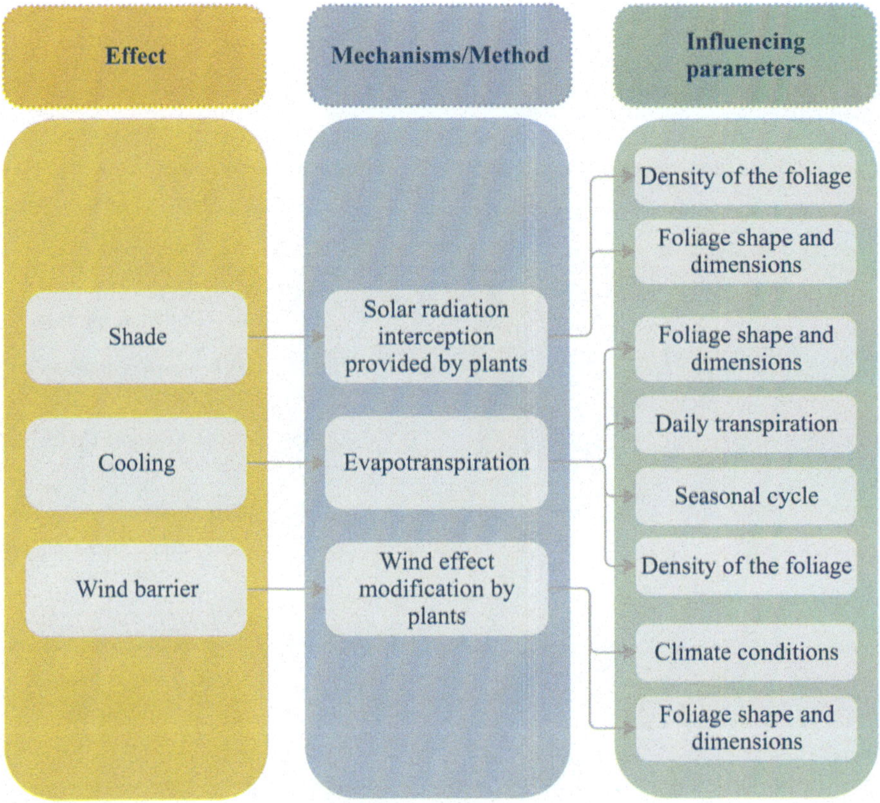

Fig. 2.10 Most relevant effects and related mechanisms/methods and influencing parameters to be considered for Nature-based solutions design based on Scudo and Ochoa de la Torre (2003), Grosso (2012) and Perini et al. (2018a)

the quantity and quality of stormwater runoff through the use of various Nature-based solutions (more specifically, Green Blue Infrastructure (GBI) systems) (Melbourne Water 2016). A treatment train integrates multiple treatment methods to improve water management outcomes at different scales. The treatment train consists of four stages (SuDS Wales 2016): (1) prevention, (2) source control, (3) site control, and (4) regional control. Water moves from one stage to the next (conveyance), passing through various systems such as pipes, swales, and linear wetlands (SuDS Wales 2016).

The first step, prevention, focuses on site design for buildings and open spaces to minimize runoff and pollution, as well as techniques for collecting and reusing stormwater. All nature-based solutions can contribute to this step. The second step, source control, addresses the management of runoff components through elements like green roofs, tree groups, meadows, and more. The third stage, local control, involves infiltration or detention, which in urban areas can be achieved through rain

Fig. 2.11 Most influencing parameters for NbS built structures based on Coma et al. (2018) and Pérez and Coma (2018)

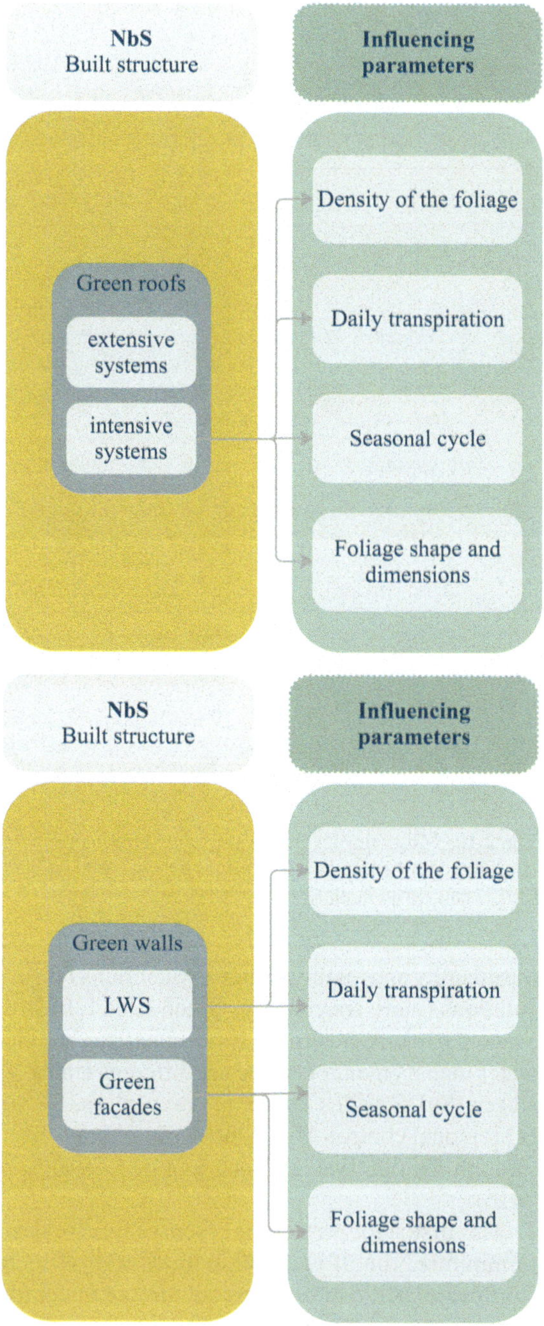

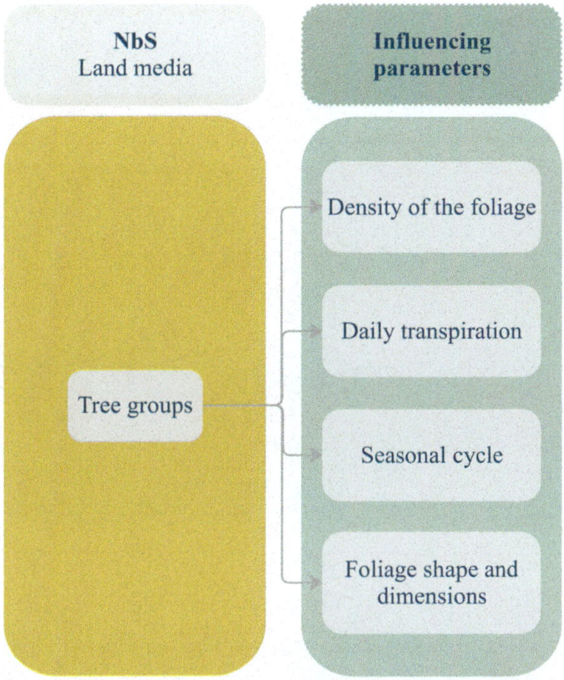

Fig. 2.12 Most influencing parameters for NbS land media (based on Scudo and Ochoa de la Torre 2003; Grosso 2012; Perini et al. 2018a)

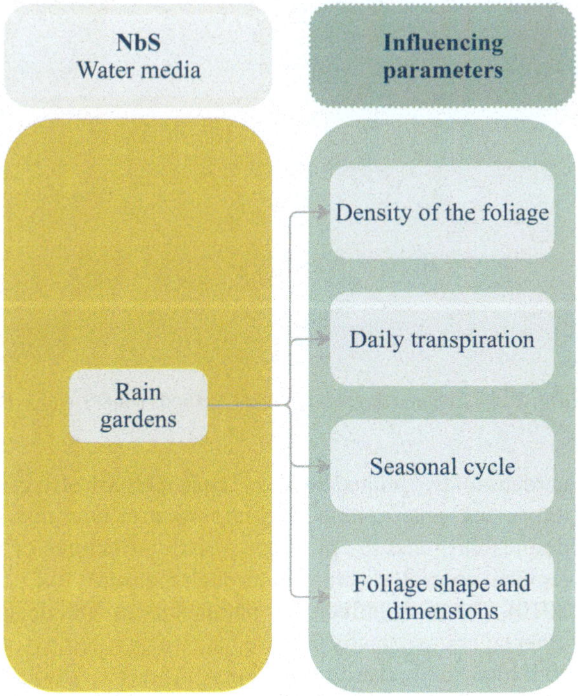

Fig. 2.13 Most influencing parameters for NbS water media based on Godwin et al. (2011) and Magliocco et al. (2020a, b)

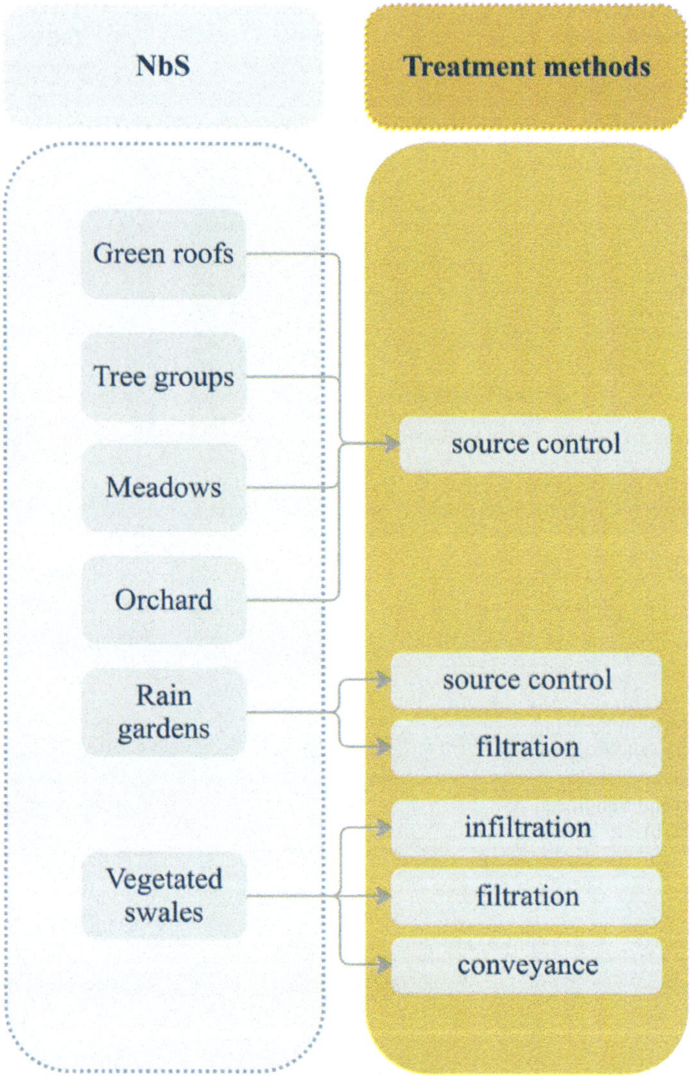

Fig. 2.14 NbS and related treatment methods (SuDS Wales 2016; Magliocco et al. 2020a, b)

gardens and vegetated swales. These NbS are also crucial for filtering rainwater and improving water quality. It's important to note that infiltration depends on ground permeability and the presence of built structures or infrastructure. The final stage, regional control, is achieved through measures like ponds and wetlands (Dickie et al. 2010). The most influencing parameters in NbS design are depth of soil, gravel and other layers, especially for to reduce water volume (source control), water silt removal (filtration) and soak through subsoil layer (infiltration). In addition, the vegetation

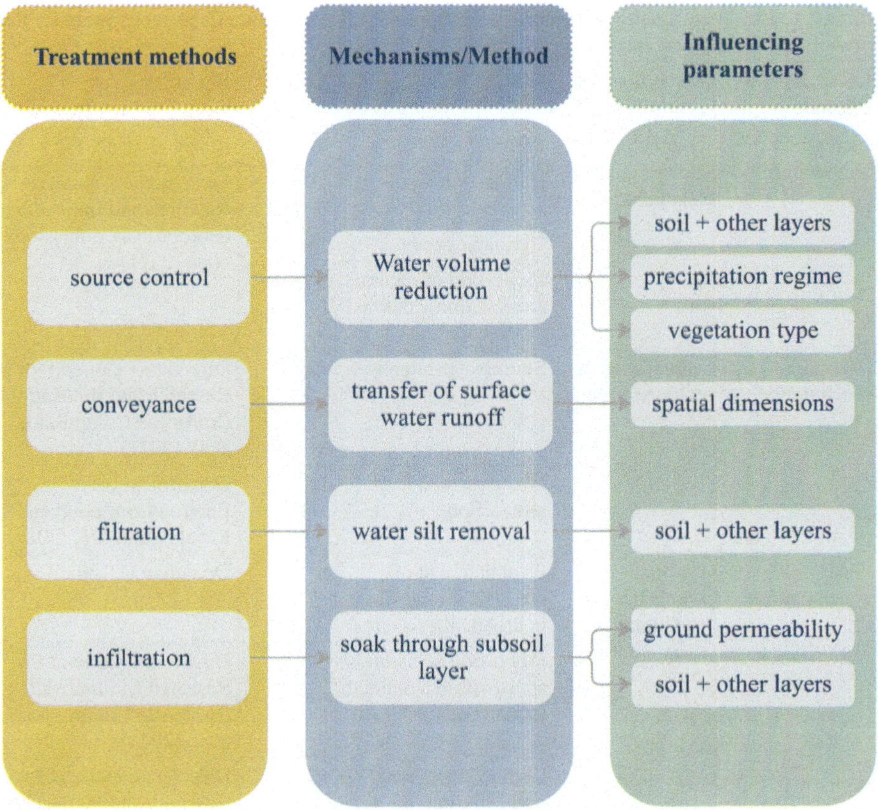

Fig. 2.15 Treatment methods, mechanism/methods and related most influencing parameters (SuDS Wales 2016; Magliocco et al. 2020a, b)

type affects water volume reduction (SuDS Wales 2016; Magliocco et al. 2020a, b). Anyhow, NbS performance deeply depend on precipitation regimes (Fioretti et al. 2010) (Fig. 2.15).

2.4 Key Performance Indicators

The performance of nature-based solutions (NbS) for climate change adaptation can be assessed using specific Key Performance Indicators (KPIs, Table 2.1). The listed KPIs, in relation to climate change drivers and effects and related environmental risks, as well as the associated regulating ecosystem services, span various scales, from the building to the district/city level, and can be measured using a range of calculation methodologies. The climate change drivers and effects increased temperature, altered precipitation, and climate change drivers and effects.

Table 2.1 Key Performance Indicators in relation to CC drivers and effects, ecosystem services provided by NbS, based on Mosca et al. (2023)

CC drivers and effects	Regulating ES	KPI(s)	Unit	References
Increased temperature	Local climate and air quality regulation	Thermal insulation	W/m^2K	Mazor et al. (2011)
		Human comfort	°C (UTCI)	Directorate-General for Research and Innovation (European Commission) et al. (2021)
		UHI incidence	%	
		Proportion of natural areas within a defined urban zone	%	
Altered precipitation	Moderation of extreme events	Stormwater retention	L	Directorate-General for Research and Innovation (European Commission) et al. (2021)
		Water use and abstraction	L	Department for Environment, Food and Rural Affairs UK (2011)
GHG concentration	Carbon sequestration and storage	Establishment of green roofs and trees in urban areas	n	Donatti et al. (2020)
		Proportion of natural areas within a defined urban zone	%	Directorate-General for Research and Innovation (European Commission) et al. (2021)
		Carbon storage and sequestration in vegetation	g	Directorate-General for Research and Innovation (European Commission) et al. (2021)
		Total carbon removed or stored in vegetation and soil per unit area per unit time	kg/ha/y	Directorate-General for Research and Innovation (European Commission) et al. (2021)

GHG concentration are mainly associated with the environmental risks extreme weather events, critical change to Earth systems, and pollution (Mosca et al. 2023).

In terms of temperature increase, NbS can contribute by regulating the local climate (Perini et al. 2018b). More specifically, built structure NbS can enhance thermal insulation and reduce cooling and heating energy consumption, as demonstrated primarily by field measurements (Coma et al. 2016, 2017). On the other hand, human comfort can be effectively modeled computationally at the local scale (see Chaps. 3 and 5). In this context, the most influential NbS are land-based media, as discussed in Sect. 2.2. The impact of urban heat islands is more significant at larger scales, such as the district level. Therefore, to mitigate such effect, a wide range of urban surfaces has to be covered by nature based solutions. This performance can also be assessed by measuring the proportion of natural areas within a defined urban zone

(Directorate-General for Research and Innovation (European Commission) et al. 2021).

To assess the performance of NbS in moderating extreme events, in relation to the climate change drivers and effects such as altered precipitation, the most relevant KPIs identified are stormwater retention and water use and abstraction. This performance is relevant at both the local and district scales and can be assessed through computational modeling (as discussed in Chaps. 4 and 5) as well as field measurements (Jayasooriya and Ng 2014).

Finally, the ecosystem service of carbon sequestration and storage can be assessed through simple calculations, such as measuring the green surface area resulting from the establishment of green roofs and trees in urban areas, or by determining the proportion of natural areas within a defined urban zone. Carbon storage and sequestration in vegetation, as well as the total carbon removed or stored in vegetation and soil per unit area per unit time (which depends on the amount of biomass) (Baggaley et al. 2022), are assessed through modeling considering the species introduced in green areas and their biological features such as dimension and wood density (Fares et al. 2017).

2.5 Summary

This chapter describes the main performances of nature-based solutions (NbS), highlighting the role they play in climate change adaptation through built structures, land media, and water media solutions. By outlining the ecosystem services provided by nature in cities, it allows for a discussion on the potential contributions of NbS to achieving relevant Sustainable Development Goals (SDGs), such as promoting sustainable cities and communities (SDG 11), enhancing resilience and adaptive capacity to climate-related disasters (SDG 13), and protecting human health (SDG 3). Additionally, by moderating extreme events, NbS also contribute to SDGs 2 (Zero Hunger), 6 (Clean Water and Sanitation), 13 (Climate Action), and 14 (Life on Land).

Given the focus on human comfort and stormwater retention, the chapter provides a detailed description of the most relevant effects, mechanisms, and methods, along with the related NbS influencing parameters. It demonstrates that performance depends not only on the type of NbS but also on factors such as plant species selection (e.g., for foliage density in microclimate regulation) and other factors, especially for stormwater retention (e.g., for infiltration).

The Key Performance Indicators (KPIs) described in the chapter can support the design and monitoring of NbS addressing climate change drivers and effects, such as increased temperature, altered precipitation patterns, and higher greenhouse gas concentrations. These indicators are also highly relevant for environmental risk management and prevention, thanks to the ecosystem services provided by NbS. The performance of built structure, land media, and water media NbS can be measured using different methodologies, such as field measurements and computational models, at various scales. In the following chapters, we will focus on computational approaches to predict the performances of NbS to inform and drive the design process to support urban resilience and adaptation to climate change.

References

Agra H, Klein T, Vasl A et al (2017) Measuring the effect of plant-community composition on carbon fixation on green roofs. Urban for Urban Green 24:1–4. https://doi.org/10.1016/j.ufug.2017.03.003

Alexandri E, Jones P (2008) Temperature decreases in an urban canyon due to green walls and green roofs in diverse climates. Build Environ 43:480–493. https://doi.org/10.1016/j.buildenv.2006.10.055

Aronson MF, Lepczyk CA, Evans KL et al (2017) Biodiversity in the city: key challenges for urban green space management. Front Ecol Environ 15:189–196. https://doi.org/10.1002/fee.1480

Babí Almenar J, Elliot T, Rugani B et al (2021) Nexus between nature-based solutions, ecosystem services and urban challenges. Land Use Policy 100:104898. https://doi.org/10.1016/j.landusepol.2020.104898

Baggaley N, Britton A, Sandison F et al (2022) Understanding carbon sequestration from nature-based solutions. https://doi.org/10.7488/ERA/2344

Baik J-J, Kwak K-H, Park S-B, Ryu Y-H (2012) Effects of building roof greening on air quality in street canyons. Atmos Environ 61:48–55. https://doi.org/10.1016/j.atmosenv.2012.06.076

Bakhshoodeh R, Ocampo C, Oldham C (2022) Evapotranspiration rates and evapotranspirative cooling of green façades under different irrigation scenarios. Energy Build 270:112223. https://doi.org/10.1016/j.enbuild.2022.112223

Ballard BW, Kellagher R, Martin P, et al (2007) Site handbook for the construction of SUDS. CIRIA

Castleton HF, Stovin V, Beck SBM, Davison JB (2010) Green roofs; building energy savings and the potential for retrofit. Energy Build 42:1582–1591. https://doi.org/10.1016/j.enbuild.2010.05.004

Chen H-S, Lin Y-C, Chiueh P-T (2023) Nexus of ecosystem service-human health-natural resources: the nature-based solutions for urban PM2.5 pollution. Sustain Cities Soc 91:104441. https://doi.org/10.1016/j.scs.2023.104441

Coma J, Pérez G, Solé C et al (2016) Thermal assessment of extensive green roofs as passive tool for energy savings in buildings. Renewable Energy 85:1106–1115. https://doi.org/10.1016/j.renene.2015.07.074

Coma J, Pérez G, de Gracia A et al (2017) Vertical greenery systems for energy savings in buildings: a comparative study between green walls and green facades. Build Environ 111:228–237. https://doi.org/10.1016/j.buildenv.2016.11.014

Coma J, Pérez G, Cabeza LF (2018) Chapter 3.2—Green roofs to enhance the thermal performance of buildings and outdoor comfort. In: Pérez G, Perini K (eds) Nature based strategies for urban and building sustainability. Butterworth-Heinemann (Elsevier), Oxford, pp 109–117

Costanza R, Mageau M (1999) What is a healthy ecosystem? Aquat Ecol 33:105–115. https://doi.org/10.1023/A:1009930313242

Department for Environment, Food and Rural Affairs UK (2011) Environmental key performance indicators: Reporting guidelines for UK business. In: GOV.UK. https://www.gov.uk/government/publications/environmental-key-performance-indicators-reporting-guidelines-for-uk-business. Accessed 29 Jan 2025

Dickie S, McKay G, Ions L, Shaffer P (2010) Planning for SuDS—making it happen

Donatti CI, Harvey CA, Hole D et al (2020) Indicators to measure the climate change adaptation outcomes of ecosystem-based adaptation. Clim Change 158:413–433. https://doi.org/10.1007/s10584-019-02565-9

Directorate-General for Research and Innovation (European Commission), Dumitru A, Wendling L (2021) Evaluating the impact of nature-based solutions: appendix of methods. Publications Office of the European Union

Eisenberg B, Chiesa C, Fischer LK et al (2022) Nature-based solutions technical handbook factsheets

European Commission. Directorate General for Climate Action (2023) EU-level technical guidance on adapting buildings to climate change: best practice guidance. Publications Office, LU

Fares S, Paoletti E, Calfapietra C et al (2017) Carbon sequestration by urban trees. In: Pearlmutter D, Calfapietra C, Samson R et al (eds) The urban forest: cultivating green infrastructure for people and the environment. Springer International Publishing, Cham, pp 31–39

Fernández-Cañero R, Pérez Urrestarazu L, Perini K (2018) Chapter 2.1—Vertical greening systems: classifications, plant species, substrates. In: Pérez G, Perini K (eds) Nature based strategies for urban and building sustainability. Butterworth-Heinemann (Elsevier), Oxford, pp 45–54

Fioretti R, Palla A, Lanza LG, Principi P (2010) Green roof energy and water related performance in the Mediterranean climate. Build Environ 45:1890–1904. https://doi.org/10.1016/j.buildenv.2010.03.001

Godwin DC, Cahill M, Sowles M (2011) Rain gardens

Grimmond S (2007) Urbanization and global environmental change: local effects of urban warming. Geogr J 173. https://doi.org/10.1111/j.1475-4959.2007.232_3.x

Grosso M (2012) La Ventilazione Naturale Controllata e il Raffrescamento Passivo Ventilativo degli edifici. http://porto.polito.it/2579961/. Accessed 2 Feb 2017

Haq A, Bakshi BR, Kodamana H, Ramteke M (2024) Assessing the effectiveness of improving urban air quality with solutions based on technology, nature and policy. Sustain Cities Soc 110:105549. https://doi.org/10.1016/j.scs.2024.105549

Hirata A, Ohashi H, Hasegawa T et al (2024) The choice of land-based climate change mitigation measures influences future global biodiversity loss. Commun Earth Environ 5:1–10. https://doi.org/10.1038/s43247-024-01433-4

IUCN (2016) Nature-based Solutions. In: IUCN. https://www.iucn.org/commissions/commission-ecosystem-management/our-work/nature-based-solutions. Accessed 20 Sep 2021

IUCN, International Union for Conservation of Nature (2020) IUCN global standard for nature-based solutions: a user-friendly framework for the verification, design and scaling up of NbS, 1st edn. IUCN, International Union for Conservation of Nature

Jayasooriya VM, Ng AWM (2014) Tools for modeling of stormwater management and economics of green infrastructure practices: a review. Water Air Soil Pollut 225:2055. https://doi.org/10.1007/s11270-014-2055-1

Köhler M (2008) Green facades—a view back and some visions. Urban Ecosyst 11:423–436. https://doi.org/10.1007/s11252-008-0063-x

Lemieux C, Bichai F, Boisjoly G (2023) Synergy between green stormwater infrastructure and active mobility: a comprehensive literature review. Sustain Cities Soc 99:104900. https://doi.org/10.1016/j.scs.2023.104900

Li M (2024) Research on the effects of extreme heat exposure on human health. Theor Nat Sci 29:194–199. https://doi.org/10.54254/2753-8818/29/20240777

Maes J, Liquete C, Teller A et al (2016a) An indicator framework for assessing ecosystem services in support of the EU biodiversity strategy to 2020. Ecosyst Serv 17:14–23. https://doi.org/10.1016/j.ecoser.2015.10.023

Maes J, Zulian G, Thijssen M et al (2016b) Mapping and assessment of ecosystems and their services—urban ecosystems 4th report

Magliocco A et al (2020a) Infrastrutture Verdi Per L'adattamento AI Cambiamenti Climatici Strategie e indicazioni progettuali per la gestione sostenibile delle acque meteoriche urbane nell'area mediterranea nord-occidentale

Magliocco A, Perini K, Sabbion P (2020b) Infrastrutture verdi per l'adattamento ai cambiamenti climatici. Infrastructures écologiques pour l'adaptation aux changements climatiques, ModusOperandi Editore

Manickathan L, Defraeye T, Allegrini J et al (2018) Parametric study of the influence of environmental factors and tree properties on the transpirative cooling effect of trees. Agric for Meteorol 248:259–274. https://doi.org/10.1016/j.agrformet.2017.10.014

Masson V, Lemonsu A, Hidalgo J, Voogt J (2020) Urban climates and climate change. Annu Rev Environ Resour 45:411–444. https://doi.org/10.1146/annurev-environ-012320-083623

Mazor MH, Mutton JD, Russell DAM, Keoleian GA (2011) Life cycle greenhouse gas emissions reduction from rigid thermal insulation use in buildings. J Ind Ecol 15:284–299. https://doi.org/10.1111/j.1530-9290.2010.00325.x

McPhearson T, Andersson E, Elmqvist T, Frantzeskaki N (2015) Resilience of and through urban ecosystem services. Ecosyst Serv 12:152–156. https://doi.org/10.1016/j.ecoser.2014.07.012

Melbourne Water (2016) Porous paving. http://www.melbournewater.com.au/planning-and-building/stormwater-management/wsud_treatments/pages/porous-paving.aspx. Accessed 11 May 2016

Morabito M, Crisci A, Gioli B et al (2015) Urban-hazard risk analysis: mapping of heat-related risks in the elderly in major Italian cities. PLoS ONE 10:e0127277

Millennium Ecosystem Assessment (Program) (ed) (2005) Ecosystems and human well-being: synthesis. Island Press, Washington, DC

Mosca F, Dotti Sani GM, Giachetta A, Perini K (2021) Nature-based solutions: thermal comfort improvement and psychological wellbeing, a case study in Genoa, Italy. Sustainability 13:11638. https://doi.org/10.3390/su132111638

Mosca F, Canepa M, Perini K (2023) Strategies for adaptation to and mitigation of climate change: key performance indicators to assess nature-based solutions performances. Urban Clim 49:101580. https://doi.org/10.1016/j.uclim.2023.101580

Oke TR (1982) The energetic basis of the urban heat Island. Quart J R Meteorol Soci 108:1–24

Pérez G, Coma J (2018) Chapter 2.3—Green roofs classifications, plant species, substrates. In: Pérez G, Perini K (eds) Nature based strategies for urban and building sustainability. Butterworth-Heinemann (Elsevier), Oxford, pp 65–74

Perini K, Sabbion P (2016) Urban sustainability and river restoration: green and blue infrastructure. Wiley

Perini K, Chokhachian A, Auer T (2018a) Chapter 3.3—Green streets to enhance outdoor comfort. In: Pérez G, Perini K (eds) Nature based strategies for urban and building sustainability. Butterworth-Heinemann (Elsevier), Oxford, pp 119–129

Perini K, Chokhachian A, Auer T (2018b) Green streets to enhance outdoor comfort. In: Nature based strategies for urban and building sustainability. Elsevier, pp 119–129

Scudo G, Ochoa de la Torre JM (2003) Spazi verdi urbani: la vegetazione come strumento di progetto per il comfort ambientale negli spazi abitati. Sistemi editoriali : Esselibri-Simone, [Napoli]

SuDS Wales (2016) The SuDS treatment train: SuDS Wales—sustainable drainage systems. http://www.sudswales.com/education/background/the-suds-treatment-train/. Accessed 5 May 2016

Taha H (1997) Urban climates and heat islands: albedo, evapotranspiration, and anthropogenic heat. Energy Build 25:99–103. https://doi.org/10.1016/S0378-7788(96)00999-1

Thompson V, Mitchell D, Hegerl GC et al (2023) The most at-risk regions in the world for high-impact heatwaves. Nat Commun 14:2152. https://doi.org/10.1038/s41467-023-37554-1

UNEA (2022) United Nations environment assembly agrees nature-based solutions definition. In: Nature-based solutions initiative. https://www.naturebasedsolutionsinitiative.org/news/united-nations-environment-assembly-nature-based-solutions-definition/. Accessed 25 Mar 2025

United Nations (2015) Sustainable development goals | United Nations Development Programme. https://www.undp.org/sustainable-development-goals/below-water?gclid=Cj0KCQjwwvilBhCFARIsADvYi7LwZm-TrGYAUA2JeQnn8ZhJ7KZms7U7MoQRNjVLgZihxwujaWLoh74aArUXEALw_wcB. Accessed 24 Jul 2023

Urban Nature Atlas (2021) Olive trees garden. https://una.city/nbs/porto/olive-trees-garden. Accessed 17 Mar 2025
VirginiaTech (2013) BMP design manual
Wood SLR, Jones SK, Johnson JA et al (2018) Distilling the role of ecosystem services in the sustainable development goals. Ecosyst Serv 29:70–82. https://doi.org/10.1016/j.ecoser.2017.10.010
World Economic Forum (2024) Global risks report 2024. In: World economic forum. https://www.weforum.org/publications/global-risks-report-2024/in-full/global-risks-2024-at-a-turning-point/. Accessed 25 Mar 2025
Zotarelli L, Dukes MD, Romero CC, et al (2024) Step by step calculation of the Penman-Monteith evapotranspiration (FAO-56 Method): AE459, 2/2010. EDIS 2010. https://doi.org/10.32473/edis-ae459-2010

Chapter 3
Nature-Based Solutions for Microclimate Regulation and Human Comfort

Abstract This chapter introduces the crucial performance of NbS in terms of microclimate regulation and thermal comfort improvement, focusing first on the natural processes and features of plants, and then on the different computational approaches currently available to measure such performances. These approaches allow for modelling and simulating the effect of NbS in urban areas to improve outdoor thermal comfort and microclimate regulation in relation to climate change impacts and environmental issues (described in Chaps. 1 and 2). More specifically, the existing tools are analyzed in depth, along with the input parameters required. Two computational approaches are then selected according to the requirements defined for their integration in a design process and tested on a typical urban canyon with different NbS.

Keywords Nature-based solutions · Urban heat island effect · Outdoor thermal comfort · Computational fluid dynamics · Real-time simulations

3.1 Urban Heat Island and Nature-Based Solutions

Worldwide cities are currently facing severe environmental degradation: in particular, extreme events driven by anthropogenic climate change are among the most critical challenges (Ebi 2022). The increase in global temperature, deriving from greenhouse gases emissions, results in several risks for cities and citizens: among them, the urban heat island (UHI) effect is one of the most relevant and dangerous (Heaviside et al. 2017). The UHI effect consists of the increase of surface temperatures in urban areas due to the urbanisation processes and the introduction of impervious materials, characterize by low albedo values (Taha 1997). In addition, the progressive reduction in the presence of vegetation in cities negatively contribute to urban microclimate regulation. The role of nature-based solutions, and specifically of vegetation, indeed, is to support the microclimate regulation through evaporation and transpiration processes, as well as with the shading capacity and the reflectance capacity of plants' leaves (Tan et al. 2021). In addition, heat waves, occurring specifically in

K. Perini and F. Mosca, *Nature-Based Cities*, Digital Innovations in Architecture, Engineering and Construction, https://doi.org/10.1007/978-3-031-94612-7_3

summer, are becoming more frequent and prolonged over time. By intensifying the UHI phenomenon, heatwaves cause the increase in heat-related risks for human and ecosystem health (Heaviside et al. 2017; Yadav et al. 2023).

Among the features of the cities with a direct impact on UHI, urban morphology is particularly relevant in terms of urban climate definition and regulation, due to the direct correlation between land use and land cover and air temperature distribution (Zhou and Chen 2018; Peng et al. 2022). Theoretically, Oke identified four main categories of parameters affecting urban climate, allowing the definition of the UHI effect (Oke 2006). These categories are (1) the urban structure, determined by the forms of the buildings, and the distribution of infrastructures; (2) the type of urban cover, which depends on the distribution of artificial or natural surfaces; (3) the urban fabric, which goes more in detail regarding the materials which define the urban fabric (e.g. roads, buildings), and the presence of natural materials, and finally (4) the urban metabolism, which considers local anthropic activities and related impacts, starting from the energy and water use, air and water pollution, extraction and use of raw materials (Oke 2006).

The relationship between these four categories can support the identification of specific local urban climate zones (LCZ). The consideration of these urban features is particularly relevant when simulating urban microclimatic conditions (Fernandes et al. 2023).

The consequences of urban heat island and heatwaves can be identified in relation to urban conditions, ecosystems, and human health. Several studies have highlighted the direct correlation between microclimatic conditions—particularly air and perceived temperatures—and mortality, especially among vulnerable groups such as children, the elderly, and people with pre-existing health conditions, who may have a compromised thermoregulatory system when exposed to thermal stress (Gasparrini et al. 2015; Mitchell et al. 2016; Royé et al. 2020; Yadav et al. 2023). Given the relevance of UHI consequences, the effects of urban design on outdoor thermal comfort should be carefully considered. Among the variety of solutions that can be implemented, nature-based solutions (NbS) allow adopting a broad range of approaches and strategies with high potential for mitigating the UHI effect and reducing related risks to living organisms and ecosystems. Specifically, NbS can reduce the impact of heatwaves by improving local environmental quality (EEA 2015), through the mitigation of the UHI at the pedestrian and the district level. In particular, the positive contribution of NbS in improving microclimatic conditions and outdoor thermal comfort is mainly related to the capacity of shading provided by vegetated elements due to their geometrical configuration (Mosca et al. 2021), as well as their radiative properties (Jones 2013), as described in Chap. 2. In addition, the high levels of albedo of leaves can support the improvement of local microclimate and reduce the local mean radiant temperature (MRT) (Abdulateef and Al-Alwan 2021). "Although the introduction of NbS has proven to be a promising strategy for reducing climate change-related risks, challenges remain. These are primarily related to the design, planning, and prediction of their site-specific performance (Cohen-Shacham et al. 2016).

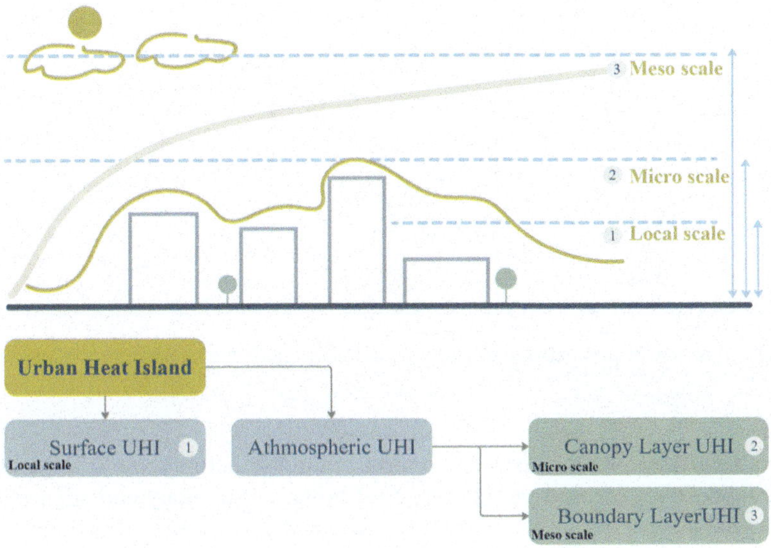

Fig. 3.1 Definition of surface-UHI, Canopy layer-UHI and urban boundary layer-UHI, based on Li et al. (2024)

The role of NbS for microclimate regulation can be identified at different scales, starting from the building scale up to the urban scale, and their effect can be correlated to the different definitions of UHI (Peng et al. 2022). In the literature, two main types of UHI, which are directly related to different scales of analysis, are defined and studied: the surface (S-UHI), and the atmospheric UHI (A-UHI), which can be subsequently divided into two main sub-categories: (2) the canopy layer (CL-UHI), and (3) the urban boundary layer (UBL-UHI, Fig. 3.1).

The main difference between the A-UHI and the S-UHI is that the increase in the local surface temperature directly defines the S-UHI. At the same time, the A-UHI characterises the atmosphere of the area considered. It is influenced by several factors, such as the urban form and materials, but also by the air flows and turbulence, which is responsible for the typical "heat dome" (Li et al. 2024).

The contribution of NbS can be identified firstly at the S-UHI level due to their capacity to provide shading and locally reduce the mean radiant temperature (MRT), as well as the transpiration processes which contribute to the reduction of local thermal stress (Abdulateef and Al-Alwan 2021). In addition, at a larger scale, the introduction of NbS such as parks and green corridors can support the reduction of the UHI effect, especially during summer heatwaves, as demonstrated by the European Spatial Agency in 2023.

Regarding this last aspect, computational modelling constitutes a promising approach due to its pre-design application and cost-effectiveness (Kumar et al. 2021).

Several computational approaches have been developed during the last years to quantify the potential of land media NbS (such as trees) and other NbS to locally

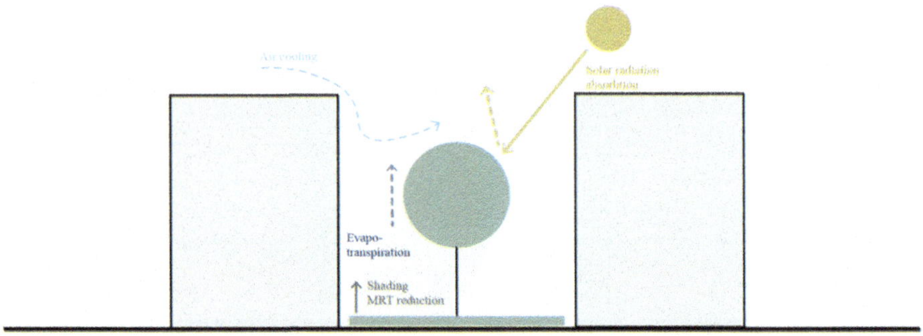

Fig. 3.2 Overview of NbS performance for microclimate regulation (adapted from Shaamala et al. 2024)

reduce the effects of heatwaves and improve outdoor thermal comfort. While most of the models are focused on the impact of urban morphology and solar radiation distribution as meaningful parameters to run real-time analysis, other more accurate approaches, among which computational fluid dynamics (CFD), have been developed during the last decades (Naboni et al. 2019), allowing a more comprehensive and detailed understanding of urban microclimate (Fig. 3.2).

In this chapter, the key computational methods and tools used to assess the effectiveness of NbS in regulating the microclimate and enhancing thermal comfort are explored. The analysis supports the identification of tools, but also the key parameters and scales of intervention. Finally, two distinct approaches, along with the relevant tools, are tested and applied to a typical urban canyon (refer to Chap. 2). The outcomes reveal the overall effectiveness of the NbS implemented and provide an overview of the advantages and disadvantages of each of the selected approach.

3.2 Computational Approaches and Tools to Quantify NbS Performances

This section provides an overview of the available modelling and simulation tools to support the quantification of NbS performance in terms of microclimate regulation. These tools can be described based on the data required, the output parameters provided, or the microclimate models used. In this chapter, two main categories are identified, depending on the type of computational approach: (1) real-time simulations and (2) computational fluid dynamics (CFD) tools. The main potential of the first category is the opportunity to run microclimatic simulations with a limited amount of time and a reduced computational power needed, on the other hand, the accuracy of these tools is limited due to the simplification introduced in the software. Conversely, the second category of tools can provide more accurate results mostly in terms of wind speed and mean radiant temperature, but requires a higher computational power and longer computing time (Mosca et al. 2025).

Below is an overview of the existing tools, divided into two categories, with a brief description of their main objectives, the types of outcomes and related data provided, and the microclimatic and comfort parameters considered (Tables 3.1 and 3.2).

Focusing on the potential use of these approaches for the simulation of NbS performance, an in-depth analysis of the modelling of plant species is needed. Concerning this aspect, the CFD approaches are considered the most appropriate (Mosca et al. 2025). In particular, among those described in Table 3.2, FLUENT introduces the "Ideal Canopy Model," which characterizes a tree based on its geometry—specifically its crown height, trunk height, and basic canopy shape, such as spherical, oval, or conical. Differently from FLUENT, OpenFOAM, PHOENICS, and ENVI-met employ a Statistical Method, linking the Leaf Area Index (LAI) or Leaf Area Density (LAD) with the plant morphology. From a mathematical perspective, concerning the calculation of plants' performance in relation to microclimate,

Table 3.1 Real-time simulation tools

Tools	Focus	Result	Parameters	References
RayMan	Deals with the effects of weather, climate, and air quality on the human organism	Estimates the radiation fluxes and the effects of clouds and solid obstacles on short-wave radiation fluxes (mean radiant temperature)	Physiological equivalent temperature (PET), predicted mean vote (PMV), standard effective temperature (SET)	Matzarakis et al. (2010)
Ladybug (GH)	Environmental analysis	Maps for solar radiation, temperature distribution, outdoor thermal comfort, etc.	PET, Universal Thermal Climate Index (UTCI)	El-Bahrawy (2024)
Dragonfly (GH)	Urban heat island modelling	Maps for temperature distribution at the urban scale	Temperature in °C	Mao and Norford (2021)
PANDO	Outdoor conditions with the possibility to set also building materials	Energy and climatic conditions of building environments, thermal and energy systems	UTCI	Chokhachian and Hiller (2020)
UMI	Urban simulation tool	Energy use of building, outdoor temperature, walkability	Dry Bulb Temperature (DBT)	Reinhart et al. (2013)
SOLWEIG	3D simulations of radiation fluxes	Urban microclimate maps including ground view factor (GVF) and well as sky view factor (SVF)	Mean radiant temperature (MRT), PET, UTCI	Lindberg et al. (2008)

Table 3.2 Computational fluid-dynamics simulation tools

Tools	Focus	Result	Parameters	References
ENVI-met (alone or combined with GH)	Simulate local microclimate considering also vegetation effects	Microclimatic fluxes	PET, UTCI	Elraouf et al. (2022)
Butterfly (GH)	CFD simulations using open foam (combined with ladybug and honeybee)	Indoor and outdoor thermal comfort and airflows	UTCI	Nguyen and Peters (2020)
OpenFOAM	CFD simulations	Maps for distribution of analysed parameters	PMV, predicted percentage of dissatisfied (PPD), and draught rate (DR) based on DIN ISO EN 7730:2005	Jasak (2009)
FLUENT	CFD simulations	Indoor simulations (maps)	PMV, PPD	Adelia et al. (2020)
FLAIR/ PHOENICS	CFD simulations	Airflow and thermal-simulation facility for building	PMV, PPD	Sun et al. (2022)
Solene coupled with Code_ Saturne code	Urban weather model implemented with a CFD code	Shadow, thermal comfort, solar energy	PET, UTCI, PMV, PPD	Morille et al. (2015)

these tools consider plants as porous media to be calculated and support the prediction of their aerodynamics.

Finally, Solene coupled with Code_Saturne allows the calculation of short-wave and long-wave radiation and its use of the evaluation of green envelopes performance is demonstrated (Morille et al. 2015).

3.3 Relevant Thermal Comfort Parameters and Scales

A wide range of parameters can be identified through a literature review regarding the performance of NbS in relation to their shape. Additionally, evaporation and transpiration processes, which contribute positively to microclimate regulation and the improvement of thermal comfort for humans, are considered key relevant parameters (see Chap. 2). Below, a description of the most relevant parameters for thermal comfort simulation is given (Blazejczyk et al. 2012; Di Napoli et al. 2018; Asghari et al. 2019; Aleksandrowicz et al. 2023).

(1) *Air temperature (°C)*: Air temperature constitutes one of the key parameters for thermal comfort assessment. It can be calculated through thermometers, and it quantifies the temperature of local air. From a technical perspective, according to the ASHRAE definition, air temperature is defined as the dry-bulb temperature of the area surrounding the occupant (Huang 2007). In addition, it is strictly related to solar radiation because atmospheric aerosol particles absorb part of the solar radiation causing the increase of atmosphere temperature (NASA 2023).
(2) *Mean radiant temperature (MRT, °C)*: This parameter is central to understanding the radiant heat exchange between the outdoor environment and the human body. The importance of MRT has grown significantly in recent decades, both due to the rise of radiant heating and cooling systems and the increasing interest in the impact of outdoor solar irradiance. The relevance of MRT is demonstrated by a wide number of scientific works, however, calculation methods for their measurement and prediction are particularly complex (Guo et al. 2020).

Concerning the role of NbS in controlling MRT, the shading contribution of vegetated systems can effectively reduce local MRT values and subsequently increase thermal comfort conditions for humans, and can vary depending on the foliage density and the shape of the solution (Coma et al. 2018; Makido et al. 2019; Perini et al. 2022).

(3) *Wind Speed and direction (m/s)*: NbS spatial distribution in cities can positively improve thermal comfort for humans by modifying wind speed and distribution. In particular, the cooling effect of NbS can be optimized by locating groups of trees or other vegetated systems following the main direction of local wind (Xi et al. 2022). The optimization of Wind Speed and direction is one of the key aspects for urban microclimate regulation.
(4) *Relative Humidity (%)*: Relative Humidity can be influenced by the evaporation processes of natural features such as trees or shrubs, or other vegetated systems. In particular, these NbS can increase local relative humidity and subsequently support the improvement of thermal comfort conditions. Compared to the other parameters listed above, relative humidity (RH) have a lower impact on perceived temperature, however, its contribution has to be considered (Hayes et al. 2022).

In addition to the microclimatic parameters, several thermal comfort indicators have been defined over the past decades. These indices relate the aforementioned parameters to predict the thermal behavior of both indoor and outdoor built and non-built spaces. Below is a brief description of two of the most commonly used indicators for outdoor thermal comfort studies, as described in the literature (Blazejczyk et al. 2012; Matzarakis et al. 2015; Asghari et al. 2019).

Physiological Equivalent Temperature (°C): This parameter has been introduced with a new scale of assessment by Höppe (Asghari et al. 2019). In particular, PET is the air temperature at which the energy budget of the human body is balanced, assuming no wind or solar radiation, while still achieving the same core and skin temperature as under actual outdoor conditions.

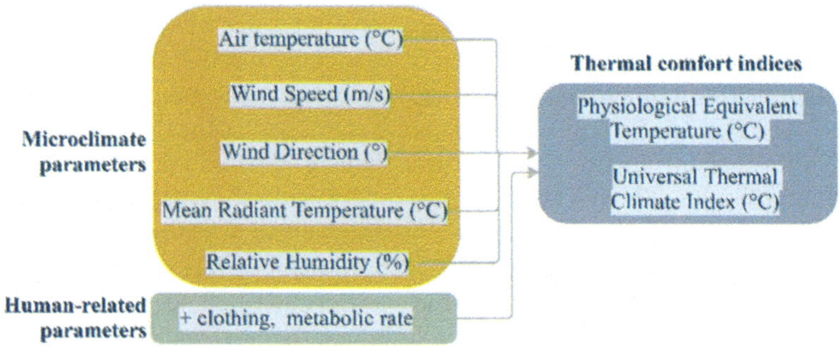

Fig. 3.3 Relevant parameters for microclimate regulation and the related impact of NbS (microclimate and human-related parameters)

Universal Thermal Climate Index (°C): The UTCI is one of the most recently developed indices for outdoor thermal comfort evaluation. It is based on thermophysiological modelling of human response to meteorological conditions. The UTCI is the temperature perceived by a specific type of human. Indeed, the UTCI considers radiant temperature, relative humidity, wind speed, and air temperature, but its calculation is also based on the individual's age and clothing (Di Napoli et al. 2018).

Additionally, another group of parameters that should be considered in the calculation of outdoor thermal comfort includes those related to physiological and behavioral aspects of humans, such as clothing, metabolic rate, and moisture permeability. While these parameters are relevant individually, they all contribute to the overall definition of thermal comfort. They are considered as inputs in a human energy balance model, which defines a temperature value that indicates the heat stress perceived by the human body. Both the temperature value and the model are generally comparable for middle European climate conditions (Perini et al. 2017, Fig. 3.3).

Typically, the calculation of thermal comfort indices through computational simulations uses default values for these parameters (e.g., 0.80 clo for clothing). However, these values can be adjusted to assess the impact of the microclimate on specific groups, such as elderly individuals.

3.4 Thermal Comfort Simulations

Based on the review of tools provided in Sect. 3.2, a general case study is presented to offer an overview of the potential applications of the most relevant and widely used tools in architectural and urban design.

In particular, two Computational Fluid Dynamics (CFD) simulation approaches are showcased. These approaches and the associated tools were selected for their ability to achieve a higher level of detail compared to other tools (Mosca et al. 2025). Based on the review and the analysis of the input required and the output provided,

ENVI-met and OpenFOAM integrated in Grasshopper have been selected and tested for implementation.

CFD models usually require a higher number of inputs and a higher level of detail in the area compared to real-time simulations, however, these approaches have been selected due to the higher level of detail achievable to support the design process.

(1) The first tool selected is ENVI-met (Bruse 2000). This standalone software focuses on the simulation of NbS performances considering both their geometrical and radiative properties. A Grasshopper plug-in has been recently developed to integrate this tool in the Rhinoceros 3D environment (Di Nunzio 2024). To run the simulation, several steps are required for model preparation. First, the geographic coordinates and the related .epw file of the site need to be defined, as well as the grid size for the model accuracy. The .epw, abbreviation for EnergyPlus Weather File, includes all climatic dad required as input (i.e. air temperature, relative humidity, wind speed and direction). This information is mandatory to define the ENVI-met workspace. Then, the 3d model of the geometry can be converted into ENVI-met objects, with details about the materials used. Once the 3d model is defined, the simulation can be set up considering the day and hour of the day and the .epw file. Outdoor thermal comfort values can be directly calculated with the simulation engine and visualised as heatmaps (as shown also in Chap. 5).
(2) The second tool selected is OpenFOAM (Jasak 2009), which can be integrated in the Rhinoceros 3D environment through the use of the Butterfly plug-in, included in the Ladybug Tools suite. The tool generates a wind tunnel which constitutes the boundary area for the simulation and considers the urban fabric elements as obstacles for the distribution of fluids within the set area. In order to simulate the dynamic of outdoor microclimatic fluids in the urban environment, specific thermal boundary conditions can be included, specifically in relation to air temperature, surface temperature of geometry' surfaces, and wind speed and direction. Output of the simulations can be both visualised in the Rhino interface or with the OpenFOAM visualisation tools; however, to calculate UTCI or other thermal comfort indices, OpenFOAM needs to be coupled with other tools such as Ladybug or Honeybee (Zhang et al. 2023).

To test the selected tools and identify their potential and limitations, a general urban canyon was defined and modeled in the Rhino environment (see Chap. 2). According to the scientific literature (Zhong et al. 2016), the urban canyon (or street canyon) has been defined as the area between to regular and symmetric buildings, modelled as boxes, located on a horizontal surface. The site selected is located in Genoa, Italy, as it aligns with the study presented in Chap. 5. The selection of the site is relevant for the definition of the summer weather conditions, input required for the simulations (Fig. 3.4). The urban canyon is modelled in the 3D CAD environment through the use of Meshes, for both ground surface and building boxes.

Four configurations of urban canyon with related NbS are modelled and tested: (0) reference canyon (no NbS), (1) green roof and façade (builti-structure NbS), (2) swale (water-media NbS), (3) trees (land-media NbS)—see Fig. 2.1, Chap. 2.

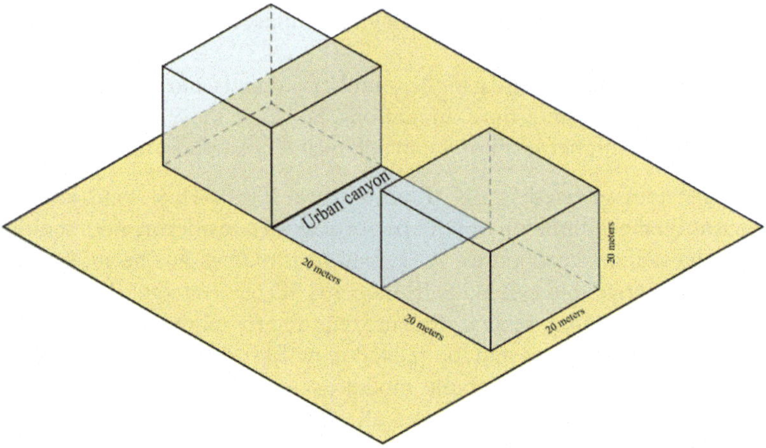

Fig. 3.4 Definition of the urban canyon used to test the selected tools and the related modelling in the CAD environment

Concerning the microclimatic parameters for the simulations, a typical summer day has been selected (21st June 2024) and the related weather conditions (Air Temperature, Wind Speed and Direction, Relative Humidity) are derived from an .epw file for the location of Genoa, a typical mediterranean city.

Table 3.3 shows the different configurations modelled to test the two simulation approaches and describes the main results in terms of UTCI values. First, the mean UTCI value for the overall area is considered, as relevant parameter to describe the level of thermal stress of the site. Secondly, specific UTCI values in the immediate surroundings of the NbS implemented (namely NbS built structure, swale and trees) are extracted, in order to showcase the potential performance of the different solutions and type of vegetation in terms of microclimate regulation.

Overall, results from the simulations show the positive contribution of land-media NbS in the urban environment for thermal stress reduction, compared to water-media and built-structure NbS, that have a lower impact on microclimate improvement (22.25 °C, compared to 22.53 and 23.08 °C, values referred to the first CFD tested approach). It is worth mentioning that, given the geometrical configuration of the urban canyon and the geographic location, the area between the two simplified buildings is already partially shaded, therefore, the performance of the NbS is influenced by the surroundings; however, the results are consistent with other studies on the topic (Hami et al. 2019; Abdi et al. 2020).

The use of the same urban canyon configuration allows the analysis of thermal comfort improvement provided by the selected NbS through the performance calculation and comparison, since the boundary conditions are consistent. Furthermore, these results demonstrate that, even in dense urban areas where buildings' shading already reduces thermal stress during the hottest hours, NbS can further enhance the livability of cities. In relation to sunlight conditions, other studies have already widely demonstrated the effectiveness of vegetation and vegetated systems in providing

Table 3.3 Simulation of the four canyon configuration and related output

Configuration	3D model	1 CFD (ENVI-met)	2 CFD (OpenFOAM)
0. Reference urban canyon	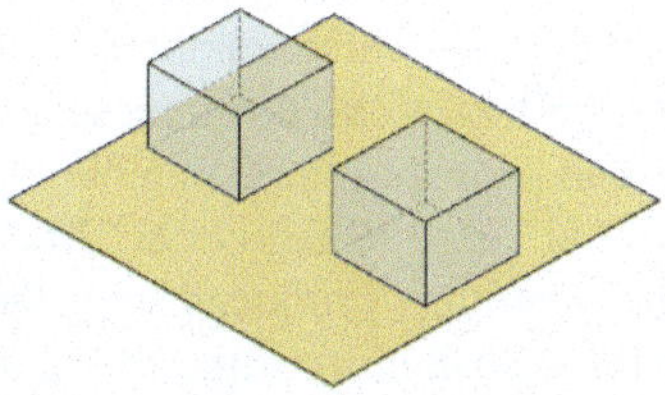	Mean UTCI: 30.64 °C UTCI in shadow area: 24.73 °C UTCI in sunlight: 31.08 °C	Mean UTCI: 27.96 °C UTCI in shadow area: 24.69 °C
1. Green roof and façade	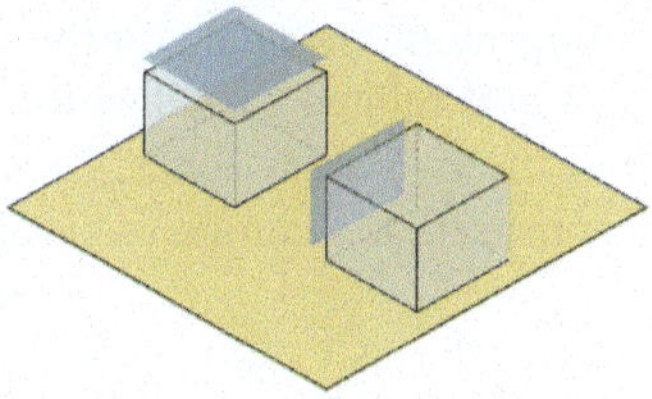	Mean UTCI: 30.49 °C UTCI in shadow area: 27.35 °C	Mean UTCI: 27.87 °C UTCI in shadow area: 24.69 °C
2. Swale	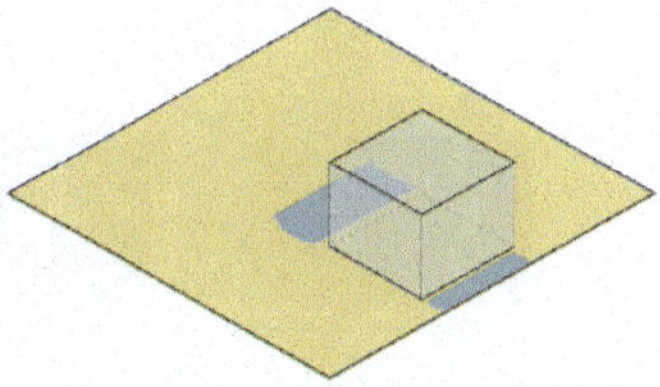	Mean UTCI: 30.03 °C UTCI near swale: 28.46 °C	Mean UTCI: 27.86 °C UTCI near swale: 27.94 °C
3. Trees	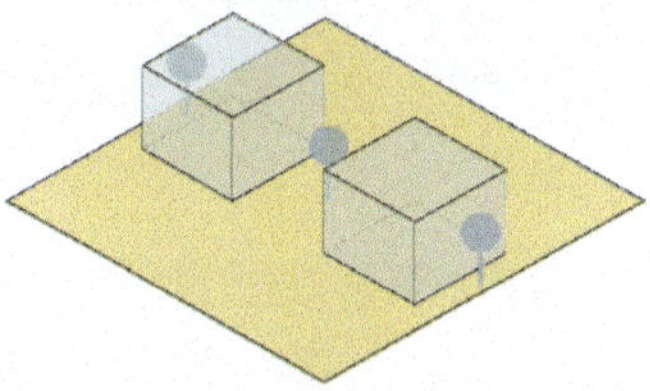	Mean UTCI: 30.13 °C UTCI under trees: 26.45 °C	Mean UTCI: 27.8 °C UTCI in shadow area: 23.95 °C

shaded areas and reducing heat-related risks for humans (Mosca et al. 2025). In particular, the contribution of the evapotranspiration process is especially significant in the case of groups of trees or shrubs (land-based NbS), in addition to the contribution from their shape (Martini et al. 2020).

Conversely, the contribution of built-structure NbS is limited due to the size and shape of the plants used (typically herbs or small shrubs in green roofs) and the scale of the solution (usually, greening solutions for building envelopes are implemented at the surface level, without substantially altering the building's shape) (Medl et al. 2017; Fernández-Cañero et al. 2018; Mayrand and Clergeau 2018). However, the role of these solutions can be effective in very dense urban areas where the introduction of land-media NbS can be difficult (Mosca et al. 2025). Finally, the main contribution of water-media NbS, and specifically NbS with permanent water bodies, in terms of microclimate regulation is provided by the volume of water, that can support the reduction of local surface temperatures, however, the effective improvement of thermal comfort conditions for humans is limited (Shimazaki et al. 2021; Cheng et al. 2022). Concerning the overall role of water for human comfort, it is worth mentioning that the presence of water bodies, with a different scale and role compared to water-based NbS in urban areas can provide interesting benefits also in terms of psychological well-being (Coburn et al. 2019; Spano et al. 2021). Figures 3.5, 3.6 and 3.7 provide an overview of the heatmaps produced with the two tested tools, offering a visual representation of the distribution of the thermal comfort indicator on the different urban canyon configurations. The results show the role played by buildings and by land-media NbS in improving thermal comfort.

The overview of simulation results and related visualization allows the identification of potentials and limitation of the two approaches, but also the discussion of some common results. An initial consideration can be focused on the simulation

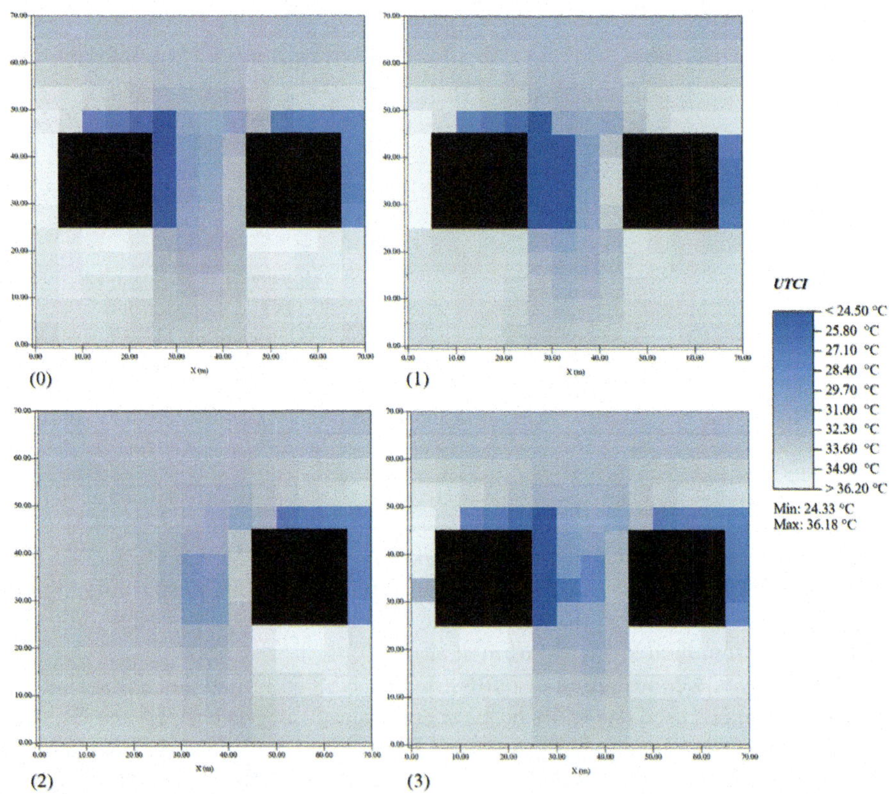

Fig. 3.5 Heatmaps with horizontal sections for the four configurations of canyon with NbS and related UTCI values from the first CFD approach tested (ENVI-met)

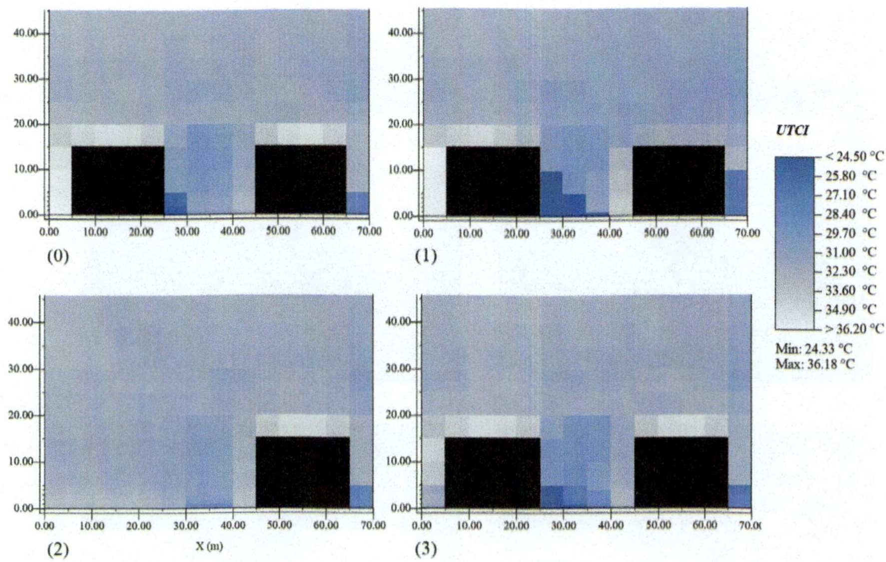

Fig. 3.6 Heatmaps with vertical sections for the four configurations of canyon with NbS and related UTCI values from the first CFD approach tested (ENVI-met)

of water-based NbS: in particular, the simulation of water features, and their potential impact on the perceived temperature, is possible only with the ENVI-met tool, while the other tested tool does not include the simulation of water features. Therefore, while planning NbS for both microclimate regulation and stormwater management, the use of the first tool could be particularly relevant. Another key aspect, when calculating microclimate conditions, is related to the calculation of MRT (as described in Sect. 3.2): the comparison of the simulation outcomes has highlighted a relevant difference between the two approaches. Specifically, ENVI-met provides higher MRT values compared to OpenFOAM, while using the same input weather data. This aspect results particularly relevant when modelling hot or arid geographical areas, where the influence of MRT on thermal comfort is crucial (Alzahrani and Gadi 2024; Sadakorn et al. 2025). Finally, another aspect to be discussed is the modelling of air flows. Specifically, the second tool tested, OpenFOAM, through the definition of a wind tunnel and the discretization of the buildings and NbS meshes, simulates with a high level of detail the distribution of wind in the canyon, that can have a meaningful impact on perceived thermal conditions (Liu et al. 2022; Ricci 2024).

Starting from the results of the two tested approaches, the following conclusions can be drafted:

(1) ENVI-met allows simulating of a wide range of well-established NbS, focusing not only on the type of NbS but also on the plant species and related technical features, such as the plant dimensions and radiative properties. This

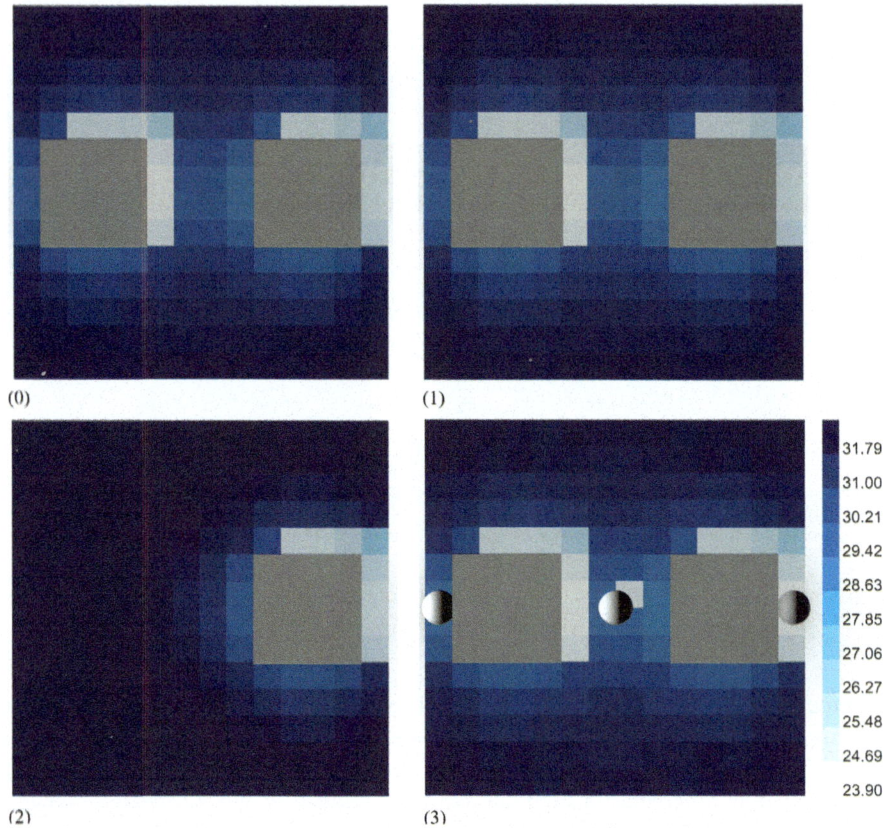

Fig. 3.7 Heatmaps with horizontal sections for the four configurations of canyon with NbS and related UTCI values from the second CFD approach tested (OpenFOAM)

feature allows designers to simulate not only the shape effect of NbS on microclimate, but also their evaporation capacity and related performance. When designing NbS with a focus on microclimate performance but considering also the ecological impact of the design, the use of this tool would be recommended.

On the other hand, through validation activities carried out with field measurement on real case studies, some studies have demonstrated the limitation of this tool in terms of the calculation of Wind Speed, which could be underestimated, depending on the level of detail of the model and on the weather conditions (Elraouf et al. 2022; Aleksandrowicz et al. 2023; Liu et al. 2024). Other studies have also highlighted the potential underestimation of MRT conditions, which is however more accurate compared to other available tools (Mosca et al. 2025). Finally, the discretization of the 3D model from the computer aided design (CAD) environment into the ENVI-met interface can greatly impact on the resolution and level of detail of the model.

(2) OpenFOAM produces more accurate results in terms of wind speed and distribution in the urban canyon, based on a well-established CFD engine. The tool reproduces a wind tunnel and calculates the distribution of the fluids based on the geometry of the obstacles modelled, i.e. buildings and NbS. Therefore, the level of detail achievable with building geometries is higher compared to the other tested tool, due to the simulation of meshes in a theoretical wind tunnel, instead of voxelized goemetries. On the other hand, in order to calculate outdoor microclimate conditions and related thermal comfort indexes, such as the MRT and the UTCI, wind speed and air temperature calculated with the CFD approach must be integrated in a real-time simulation engine (e.g. Ladybug tools), characterized by limitations in terms of MRT calculation, or other CFD approaches.

Overall, the application of the two methodologies the same urban canyon with different NbS have demonstrated their potential in quantifying outdoor thermal comfort conditions in urban areas to support the decision-making process both at early design stages to optimize the distribution of NbS in outdoor spaces, and as a final assessment. Depending on the peculiarities of the site considered, and the specific requirements of design, a specific tool can be selected. For example, in very dense urban areas, where calculating the distribution of air flows can be particularly difficult, OpenFOAM can be selected; on the other hand, when a site is characterized by a very high solar exposure, and the calculation of MRT constitutes the main challenge, ENVI-met should be considered for its capacity to simulate the distribution of this key parameter.

3.5 Summary

This chapter provides a comprehensive overview of the role of different types of NbS in microclimate regulation and outdoor thermal comfort improvement in the urban environment. The most relevant parameters and tools for the quantification of this key performance have been identified and described, and a series of applications on the same urban canyon with different NbS is provided, in order to allow the identification of the trade-offs between the different available approaches. Among the different tools available, two computational fluid dynamics tools have been selected, due to their higher precision. By testing the two selected approaches with same design and environmental starting conditions (urban canyon configurations, geo-localization, climatic data used), namely OpenFOAM and ENVI-met, the main potentials and limitations of simulation tools for NbS performance calculation are identified and discussed. Finally, the performance of land-media, water-media, and built-structure NbS is quantified, compared, and discussed, in order to provide meaningful recommendations for their implementation in cities.

References

Abdi B, Hami A, Zarehaghi D (2020) Impact of small-scale tree planting patterns on outdoor cooling and thermal comfort. Sustain Cities Soc 56:102085

Abdulateef MF, Al-Alwan HA (2021) The effectiveness of urban green infrastructure in reducing surface urban heat island. Ain Shams Eng J

Adelia AS, Nevat I, Acero JA et al (2020) Tool comparison for urban microclimate modelling. ETH Zurich

Aleksandrowicz O, Saroglou T, Pearlmutter D (2023) Evaluation of summer mean radiant temperature simulation in ENVI-met in a hot Mediterranean climate. Build Environ 245:110881. https://doi.org/10.1016/j.buildenv.2023.110881

Alzahrani A, Gadi M (2024) Assessment of modeled mean radiant temperature in hot and dry environments: a case study in Saudi Arabia. Climate 12:91. https://doi.org/10.3390/cli12070091

Asghari M, Teimori G, Abbasinia M et al (2019) Thermal discomfort analysis using UTCI and MEMI (PET and PMV) in outdoor environments: case study of two climates in Iran (Arak & Bandar Abbas). Weather 74. https://doi.org/10.1002/wea.3612

Blazejczyk K, Epstein Y, Jendritzky G et al (2012) Comparison of UTCI to selected thermal indices. Int J Biometeorol 56:515–535. https://doi.org/10.1007/s00484-011-0453-2

Bruse M (2000) Assessing thermal comfort in urban environments using a integrated dynamic microscale biometeorological model system. https://ams.confex.com/ams/AugDavis/techprogram/paper_15005.htm. Accessed 24 Mar 2025

Cheng Y, Liu X, Zeng Z et al (2022) Impacts of water bodies on microclimates and outdoor thermal comfort: implications for sustainable rural revitalization. Front Environ Sci 10:940482. https://doi.org/10.3389/fenvs.2022.940482

Chokhachian A, Hiller M (2020) PANDO: parametric tool for simulating soil-plant atmosphere of tree canopies in grasshopper. Symposium on simulation for architecture and urban design (SimAUD), pp 495–502

Coburn A, Kardan O, Kotabe H et al (2019) Psychological responses to natural patterns in architecture. J Environ Psychol 62:133–145. https://doi.org/10.1016/j.jenvp.2019.02.007

Cohen-Shacham E, Walters G, Janzen C, Maginnis S (eds) (2016) Nature-based solutions to address global societal challenges. IUCN International Union for Conservation of Nature

Coma J, Pérez G, Cabeza LF (2018) Green roofs to enhance the thermal performance of buildings and outdoor comfort. In: Nature based strategies for urban and building sustainability. Elsevier, pp 109–117

Di Napoli C, Pappenberger F, Cloke HL (2018) Assessing heat-related health risk in Europe via the Universal Thermal Climate Index (UTCI). Int J Biometeorol 62:1155–1165

Di Nunzio (2024) Online available at: https://github.com/AntonelloDN/Morpho

Ebi KL (2022) Managing climate change risks is imperative for human health. Nat Rev Nephrol 18:74–75. https://doi.org/10.1038/s41581-021-00523-2

EEA (2015) Green infrastructure: better living through nature-based solutions—European Environment Agency

El-Bahrawy AM (2024) Investigating Ladybug as a tool for measuring outdoor thermal comfort in urban neighborhoods. J Eng Sci 52:33–57. https://doi.org/10.21608/jesaun.2023.234735.1256

Elraouf RA, ELMokadem A, Megahed N et al (2022) Evaluating urban outdoor thermal comfort: a validation of ENVI-met simulation through field measurement. J Build Perform Simul 15:268–286. https://doi.org/10.1080/19401493.2022.2046165

European Spatial Agency (2023) Heatwave across Europe. https://www.esa.int/ESA_Multimedia/Images/2023/07/Heatwave_across_Europe. Accessed 9 Oct 2024

Fernandes R, Nascimento V, Freitas M, Ometto J (2023) Local climate zones to identify surface urban heat Islands: a systematic review. Remote Sens 15:884. https://doi.org/10.3390/rs15040884

Fernández-Cañero R, Pérez Urrestarazu L, Perini K (2018) Vertical greening systems. In: Nature based strategies for urban and building sustainability. Elsevier, pp 45–54

Gasparrini A, Guo Y, Hashizume M et al (2015) Mortality risk attributable to high and low ambient temperature: a multicountry observational study. Lancet 386:369–375

Guo H, Aviv D, Loyola M et al (2020) On the understanding of the mean radiant temperature within both the indoor and outdoor environment, a critical review. Renew Sustain Energy Rev 117:109207. https://doi.org/10.1016/j.rser.2019.06.014

Hami A, Abdi B, Zarehaghi D, Maulan SB (2019) Assessing the thermal comfort effects of green spaces: a systematic review of methods, parameters, and plants' attributes. Sustain Cities Soc 49:101634. https://doi.org/10.1016/j.scs.2019.101634

Hayes A, Jandaghian Z, Lacasse M et al (2022) Nature-based solutions (NBSs) to mitigate urban heat Island (UHI) effects in Canadian cities. Buildings 12:925. https://doi.org/10.3390/buildings12070925

Heaviside C, Macintyre H, Vardoulakis S (2017) The urban heat island: implications for health in a changing environment. Curr Environ Health Rep 4:296–305

Huang J (2007) Prediction of air temperature for thermal comfort of people in outdoor environments. Int J Biometeorol 51:375–382. https://doi.org/10.1007/s00484-006-0083-2

Jasak H (2009) OpenFOAM: open source CFD in research and industry. Int J Naval Arch Ocean Eng 1:89–94. https://doi.org/10.2478/IJNAOE-2013-0011

Jones HG (2013) Plants and microclimate: a quantitative approach to environmental plant physiology, 3rd edn. Cambridge University Press

Kumar P, Debele SE, Sahani J et al (2021) An overview of monitoring methods for assessing the performance of nature-based solutions against natural hazards. Earth Sci Rev 217:103603. https://doi.org/10.1016/j.earscirev.2021.103603

Li Y, Yang T, Zhao G et al (2024) A systematic review of studies involving canopy layer urban heat island: Monitoring and associated factors. Ecol Ind 158:111424. https://doi.org/10.1016/j.ecolind.2023.111424

Lindberg F, Holmer B, Thorsson S (2008) SOLWEIG 1.0—modelling spatial variations of 3D radiant fluxes and mean radiant temperature in complex urban settings. Int J Biometeorol 52:697–713. https://doi.org/10.1007/s00484-008-0162-7

Liu K, Lian Z, Dai X, Lai D (2022) Comparing the effects of sun and wind on outdoor thermal comfort: a case study based on longitudinal subject tests in cold climate region. Sci Total Environ 825:154009. https://doi.org/10.1016/j.scitotenv.2022.154009

Liu S, Middel A, Fang X, Wu R (2024) ENVI-met model performance evaluation for courtyard simulations in hot-humid climates. Urban Clim 55:101909. https://doi.org/10.1016/j.uclim.2024.101909

Makido Y, Hellman D, Shandas V (2019) Nature-based designs to mitigate urban heat: the efficacy of green infrastructure treatments in Portland, Oregon. Atmosphere 10:282. https://doi.org/10.3390/atmos10050282

Mao J, Norford LK (2021) Urban weather generator: physics-based microclimate simulation for performance-oriented urban planning. In: Palme M, Salvati A (eds) Urban microclimate modelling for comfort and energy studies. Springer International Publishing, Cham, pp 241–263

Martini A, Biondi D, Batista AC (2020) Thermal comfort provided by street trees in cities. Arboricult J 42:153–164

Matzarakis A, Muthers S, Rutz F (2015) Application and comparison of UTCI and PET in temperate climate conditions. Finisterra 49(98). https://doi.org/10.18055/FINIS6453

Matzarakis A, Rutz F, Mayer H (2010) Modelling radiation fluxes in simple and complex environments: basics of the RayMan model. Int J Biometeorol 54:131–139. https://doi.org/10.1007/s00484-009-0261-0

Mayrand F, Clergeau P (2018) Green roofs and green walls for biodiversity conservation: a contribution to urban connectivity? Sustainability 10:985. https://doi.org/10.3390/su10040985

Medl A, Stangl R, Florineth F (2017) Vertical greening systems—a review on recent technologies and research advancement. Build Environ 125:227–239

Mitchell D, Heaviside C, Vardoulakis S et al (2016) Attributing human mortality during extreme heat waves to anthropogenic climate change. Environ Res Lett 11:074006

Morille B, Lauzet N, Musy M (2015) SOLENE-microclimate: a tool to evaluate envelopes efficiency on energy consumption at district scale. Energy Proc 78:1165–1170. https://doi.org/10.1016/j.egypro.2015.11.088

Mosca F, Calbi M, Roccotiello E, Perini K (2025) A computational approach to assess the effects of ecological building envelopes on outdoor thermal comfort. Sustain Cities Soc 106170. https://doi.org/10.1016/j.scs.2025.106170

Mosca F, Dotti Sani GM, Giachetta A, Perini K (2021) Nature-based solutions: thermal comfort improvement and psychological wellbeing, a case study in Genoa, Italy. Sustainability 13:11638. https://doi.org/10.3390/su132111638

Naboni E, Meloni M, Makey C, Kaempf J (2019) The simulation of mean radiant temperature in outdoor conditions: a review of software tools capabilities. Rome, Italy, pp 3234–3241

NASA (2023) The Earth's radiation budget—NASA Science. https://science.nasa.gov/ems/13_radiationbudget/. Accessed 30 Sep 2024

Nguyen JN, Peters B (2020) Computational fluid dynamics in building design practice. In: ACADIA proceedings 2020

Oke TR (2006) Towards better scientific communication in urban climate. Theor Appl Climatol 84:179–190. https://doi.org/10.1007/s00704-005-0153-0

Peng W, Wang R, Duan J et al (2022) Surface and canopy urban heat islands: does urban morphology result in the spatiotemporal differences? Urban Clim 42:101136. https://doi.org/10.1016/j.uclim.2022.101136

Perini K, Calise C, Castellari P, Roccotiello E (2022) Microclimatic and environmental improvement in a Mediterranean city through the regeneration of an area with nature-based solutions: a case study. Sustainability 14:5847. https://doi.org/10.3390/su14105847

Perini K, Chokhachian A, Dong S, Auer T (2017) Modeling and simulating urban outdoor comfort: coupling ENVI-Met and TRNSYS by grasshopper. Energy Build 152:373–384. https://doi.org/10.1016/j.enbuild.2017.07.061

Reinhart C, Dogan T, Jakubiec A, et al (2013) Umi—an urban simulation environment for building energy use, daylighting and walkability. IBPSA, pp 476–483

Ricci A (2024) Review of OpenFOAM applications in the computational wind engineering: from wind environment to wind structural engineering. Meccanica. https://doi.org/10.1007/s11012-024-01826-x

Royé D, Codesido R, Tobías A, Taracido M (2020) Heat wave intensity and daily mortality in four of the largest cities of Spain. Environ Res 182:109027. https://doi.org/10.1016/j.envres.2019.109027

Sadakorn W, Tetiranont S, Prasittisopin L, Kaewunruen S (2025) Assessing thermal comfort in hot and humid (tropical) climates: urban outdoor and semi-outdoor conditions in waiting areas of railway stations. Build Environ 267:112240. https://doi.org/10.1016/j.buildenv.2024.112240

Shaamala A, Yigitcanlar T, Nili A, Nyandega D (2024) Strategic tree placement for urban cooling: a novel optimisation approach for desired microclimate outcomes. Urban Clim 56:102084. https://doi.org/10.1016/j.uclim.2024.102084

Shimazaki Y, Aoki M, Nitta J et al (2021) Experimental determination of pedestrian thermal comfort on water-retaining pavement for UHI adaptation strategy. Atmosphere 12:127. https://doi.org/10.3390/atmos12020127

Spano G, Dadvand P, Sanesi G (2021) The benefits of nature-based solutions to psychological health. Front Psychol 12

Sun C, Lian W, Liu L et al (2022) The impact of street geometry on outdoor thermal comfort within three different urban forms in severe cold region of China. Build Environ 222:109342. https://doi.org/10.1016/j.buildenv.2022.109342

Taha H (1997) Urban climates and heat islands: albedo, evapotranspiration, and anthropogenic heat. Energy Build 25:99–103

Tan JKN, Belcher RN, Tan HTW et al (2021) The urban heat island mitigation potential of vegetation depends on local surface type and shade. Urban for Urban Green 62:127128. https://doi.org/10.1016/j.ufug.2021.127128

Xi C, Ding J, Wang J et al (2022) Nature-based solution of greenery configuration design by comprehensive benefit evaluation of microclimate environment and carbon sequestration. Energy Build 270:112264. https://doi.org/10.1016/j.enbuild.2022.112264

Yadav N, Rajendra K, Awasthi A et al (2023) Systematic exploration of heat wave impact on mortality and urban heat island: a review from 2000 to 2022. Urban Clim 51:101622. https://doi.org/10.1016/j.uclim.2023.101622

Zhang M, Tan S, Zhang C et al (2023) Assessing the impact of fractional vegetation cover on urban thermal environment: a case study of Hangzhou, China. Sustain Cities Soc 96:104663. https://doi.org/10.1016/j.scs.2023.104663

Zhong J, Cai X-M, Bloss WJ (2016) Coupling dynamics and chemistry in the air pollution modelling of street canyons: a review. Environ Pollut 214:690–704. https://doi.org/10.1016/j.envpol.2016.04.052

Zhou X, Chen H (2018) Impact of urbanization-related land use land cover changes and urban morphology changes on the urban heat island phenomenon. Sci Total Environ 635:1467–1476. https://doi.org/10.1016/j.scitotenv.2018.04.091

Chapter 4
Nature-Based Solutions for Stormwater Retention

Abstract This chapter first presents the main challenges of urban environments in terms of stormwater management and the potential role of nature-based solutions (NbS). Secondly, it highlights the potential of computational approaches to model and simulate the performance of NbS for stormwater retention, addressing the aforementioned impacts of climate change and environmental issues. An overview of existing tools is provided, along with a detailed description of the input parameters required for analysis and simulation, and the expected outcomes are illustrated. Two models are selected based on the requirements for their integration into the design process, tested, and applied to a generic urban canyon with various NbS. This allows for a discussion of the potential and limitations of the most relevant strategies selected, as well as an evaluation of the usability and suitability of the chosen tools.

Keywords Nature-based solutions · Rainwater · Permeable surfaces · Climate adaptation · Flooding risk reduction

4.1 Water Management and Nature-Based Solutions

Urbanization and climate change are negatively affecting the availability and reliability of freshwater resources (IPCC 2021). In particular, the IPCC report highlights that specific humidity near the Earth's surface has very likely increased, along with the water vapor content in the atmosphere. These changes, coupled with the global warming phenomenon, trigger a thermodynamic response that alters the distribution of relative humidity, consequently influencing the rainfall cycle (Oppenheimer et al. 2014; Douville et al. 2021).

The increase in global temperatures, due to rising greenhouse gas (GHG) emissions, is considered one of the main drivers of the alteration of precipitation. These changes are expected to become more extreme and less evenly distributed both throughout the year and geographically. Studies predict that the frequency of rainfall and associated floods will increase by approximately 20 to 30% due to climate change and urbanization processes (Madsen et al. 2014; Prosdocimi et al. 2015;

K. Perini and F. Mosca, *Nature-Based Cities*, Digital Innovations in Architecture, Engineering and Construction, https://doi.org/10.1007/978-3-031-94612-7_4

Hoerling et al. 2019; Allan et al. 2020). Findings from Allan et al. (2020) highlight that changes in atmospheric circulation, resulting from the evolution of surface temperature distribution, can influence the water cycle and its balance at the regional scale.

The water cycle and precipitation levels are directly influenced by climate change, as GHG emissions and atmospheric concentrations impact the net radiative cooling capacity. From an operational perspective, urbanization has caused significant land-use (LU) changes, replacing natural surfaces with artificial ones and reducing the proportion of permeable surfaces in urban settings. While in natural environments water management depends on factors such as land porosity, orography, and rainwater dynamics, urban areas exhibit greater complexity. Artificial surfaces, infrastructure, and industrial activities alter the water cycle, increasing local water runoff due to reduced permeability. Furthermore, diminished vegetation exacerbates this issue by reducing stormwater interception. These processes accelerate the flow of runoff into rivers and basins, heightening flood risks (Koop and van Leeuwen 2017, Fig. 4.1).

Local sewage systems, often overloaded during extreme storm events, are increasingly unreliable as standalone solutions in the context of climate change (Qin 2020). To enable effective stormwater management and reduce the risks associated with extreme events, various design and planning approaches are discussed in the literature, focusing on both stormwater management and treatment. These include Sustainable Urban Drainage Systems (SUDS), Water Sensitive Urban Design (WSUD), Low Impact Development (LID), and Sponge Cities (SC), among others (Fletcher et al. 2015). In 2015, the European Commission officially defined the term 'nature-based solutions' (NbS) to encompass both emerging and established technologies, such as those mentioned, as promising alternatives for stormwater management (Krull et al. 2015).

By incorporating soil, permeable surfaces and various vegetation types, NbS can reduce local water runoff, temporarily store rainwater, and alleviate pressure on

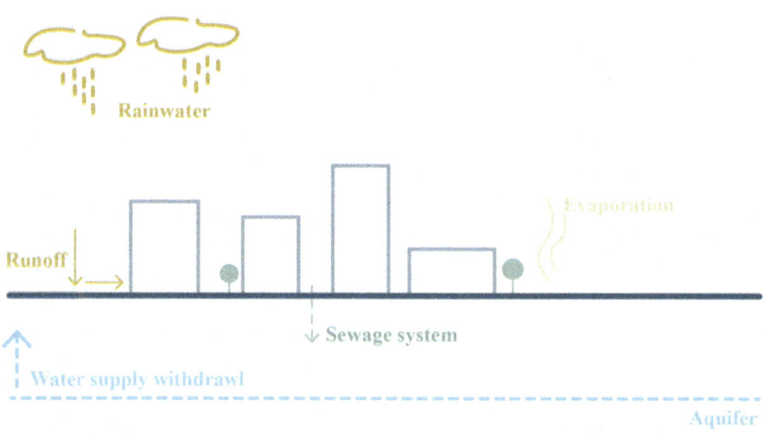

Fig. 4.1 Complexity of water cycle in the urban environment (adapted from Ridolfi 2014)

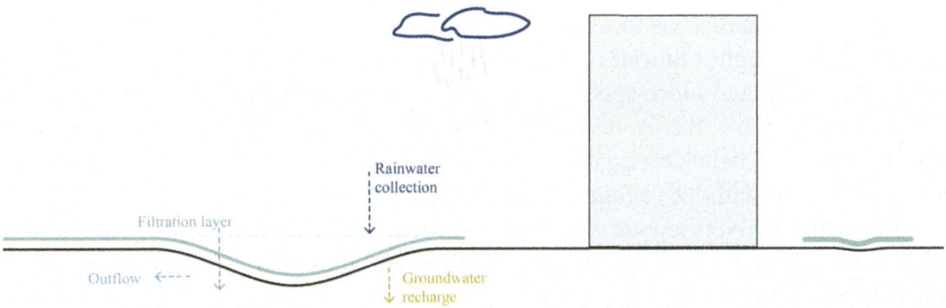

Fig. 4.2 Overview of NbS performance for stormwater retention (adapted from: Biswal et al. 2022)

existing grey infrastructures. Bioretention systems, a key category of NbS, not only mitigate rainfall-related risks but can also enhance water quality (Vijayaraghavan et al. 2021). These design and planning strategies aim to improve water management sustainably, indirectly reducing stormwater-related risks such as flooding (Qin 2020).

Water-based nature-based solutions, such as rain gardens, green roofs, infiltration channels, permeable pavements, and bio-swales, effectively mitigate rainwater runoff and reduce urban flooding risks (Palermo et al. 2023). Notably, many of these approaches are endorsed by the United States Environmental Protection Agency (US EPA 2015) as promising strategies. In addition to stormwater management, these solutions offer complementary benefits, including erosion control, water purification, and enhanced water supply (UNEP 2018).

Modeling NbS to predict their performance in stormwater retention is crucial for effective urban planning, promoting adaptation to climate change. Considering the critical importance of water as a natural resource, strategies must address both the risks posed by extreme weather events and the need to enhance rainwater collection for future use (Jayasooriya and Ng 2014).

In this chapter, we present the most relevant computational approaches and tools to quantify the performance of nature-based solutions for stormwater management in urban areas. The analysis defines the most important parameters and intervention scales. Finally, two approaches, along with their related tools, are tested and applied to a typical urban canyon with a set of different NbS (see Chap. 2). The results show the overall performance of the applied NbS and highlight the pros and cons of the different approaches (Fig. 4.2).

4.2 Computational Approaches and Tools to Quantify NbS Performances

This Section provides an overview of the available modelling and simulation tools to support the quantification of NbS performance for stormwater management. Over the past few decades, due to the growing importance of the topic and the rise in extreme events, numerous tools have been developed to quantify the impact of urban planning and design solutions on stormwater management, with a particular focus on the role of NbS, and more specifically, green–blue infrastructures. From a methodological perspective, the tools and software currently available are characterized by the definition of balance equations that allow the prediction of the behavior of a system (with or without vegetation and porous surfaces).

Below is an overview of the existing tools, including a brief description of their main objectives, the results they provide, and the input and output parameters considered (Table 4.1).

In addition, other tools are currently available to evaluate the urban stormwater quality, such as HydroWorks, MOSQUITO, or MouseTrap (Mannina and Viviani 2010). Jayasooriya and Ng 2014 provide an overview of the tools currently available to support the introduction of green infrastructures in the urban environment. The authors demonstrated that among the 20 tools identified, most of them focus on the prediction of the economic impact of GI and on the role of plants and soil in improving water quality, while just a few of them propose the quantification of the

Table 4.1 Summary of the methodologies found in the literature to evaluate the stormwater retention capacity of a system

Tools	Focus	Results	Parameters	References
Hydrus-1D (Šimůnek et al. 2008)	The model is based on a primary equation defined to evaluate the water flow in porous media	• Potential water uptake • Water content	• L • cm^3/cm^3	Zhang et al. (2021)
SWMM	The model provides the simulation for different LID solutions in relation to stormwater management. The tool is based on the Horton, Gree-Ampt, and CN methods	• Initial storage volume • Final storage volume • Flow continuity error	• m^3 • m^3 • (%)	US EPA (2014)
Open HydroQual	The tool provides a model to estimate the behavior of a water system. It is based on a general balance equation for water	Water flow	L/s)	Massoudieh et al. (2023)
Multi-hydro	The model provides a representation of the urban environment and its behaviour in relation to rainfall and related water flows. The latest implementation includes the simulation of NbS performance	Discharge of sewer system	m^3	Versini et al. (2016, 2018)

The authors have demonstrated this approach is one of the most common calculation methods available in the literature (based on a systematic literature review)

effects on water runoff, in particular: SLAMM (Pitt and Voorhees 2002), CNT Green Values National Stormwater Management Calculator (CNT 2021), SWMM (US EPA 2014), L-THIA (Engel and Theller 2016), and MUSIC (Dotto et al. 2011). However, these models mostly focus on the effects of land use on urban stormwater retention, and their application is restricted to the urban scale.

In addition, their implementation and use to support the decision-making process can be particularly challenging given the level of detail required for the simulation.

Finally, a key aspect that emerges from the analysis of existing models and software is that the scale (urban, district, or building) is particularly important for assessing performance in terms of stormwater retention. Indeed, depending on the scale and specific characteristics of a design project, the effectiveness of NbS can vary significantly, and this needs to be predicted with a focus on the actual scale being considered.

While the majority of tools available for assessing microclimate regulation and human comfort performance resulting from the implementation of NbS (Chap. 3, Sect. 3.2) are user-friendly, the tools identified through this literature review require a higher level of detail and expertise to conduct a proper hydraulic analysis of a system. This constitutes a major limitation for integrating environmental simulations into the design process. Indeed, the quantification of a design solution's performance in terms of rainwater collection is typically carried out after construction (Razguliaev et al. 2024).

However, given the increasing frequency and intensity of extreme rainfall events in urban areas worldwide, there is a need to develop more accessible and suitable tools to support the achievement of specific targets and goals related to climate adaptation and urban resilience. Identifying tools that can also be used by architects, landscape architects, and other professionals without a specific background in hydrological systems would allow for a wider use of simulations to predict and optimize the performance of NbS at a given site.

4.3 Relevant Water Management Parameters and Scales

This section provides an overview of the most relevant parameters that can be used for the quantification of the water retention capacity of an NbS, or, more in general, of an area (Soil Water Retention, an overview, 2024).

(1) *Liter of water retained (L)*: this parameter, identified as the most frequently used in literature, allows the quantification of the volume of water retained by a system. Physically, it can be considered as the volume of water itself (L), but also in relation to the surface of reference (L/m^2), or in relation to a specific timeframe, measured in hours or seconds (L/h or L/s). The last option, which calculates the retention of a specific water volume in the time unit, is specifically considered in relation to the analysis of the response of a system to extreme

events such as heavy rains and related potential floods, in order to quantify the effectiveness of the solution' response.

(2) *Soil water retention capacity (m^3/m^3)*: this parameter is commonly used for NbS, such as rain gardens, and can be used to calculate the amount of volume of water that can be retained by the total soil volume of the solution implemented. Physically, this parameter indicates the capacity of the considered soil to store water within its volume.

(3) *Hydraulic connectivity (%)*: this parameter describes the path and the deposit of water within soil fractions and pores and can be calculated as a function of the soil–water content (Maltby and Barker 2009).

(4) *Volumetric water content (cm^3/cm^3)*: this parameter can be used to predict the capacity of soil to retain rainwater. It is calculated as the division between the water volume contained in a unit of soil and the soil volume (Novák and Hlaváčiková 2019).

It is worth noting that the units adopted depend on the area or country where a tool was developed. For example, tools developed in the United States use the imperial system, which includes units such as inches, feet, yards, miles, and gallons for volume.

The list of parameters allows the identification of the main variables which need to be considered while designing to improve water management in urban relates. In particular, three main variables can be listed: (1) the volume of the different layers of NbS which can collect water during rainfalls and extreme events. This variable is crucial and should guide the planning and design process of the solutions, by defining their dimensions concerning the peak rainfall values for the selected site; (2) the soil features are crucial to define the potential retention of water within its volume; (3) the time variable is likewise important, as the duration of extreme events can influence the capacity of water retention. These parameters are particularly relevant to quantify the performance of the different NbS in relation to the main mechanisms for stormwater management illustrated in Chap. 2, namely, conveyance, filtration, and infiltration.

Finally, the impact of NbS can be considered depending on the type of solution implemented and its scale. Starting from the building scale, where NbS can be integrated on the vertical and horizontal surfaces of the buildings, the contribution can be locally quantified considering the amount of rainwater retained by the vegetation layer and the growing medium (Longobardi et al. 2019). Concerning the district and the urban scale, the contribution of these solutions can vary depending on their extension and their technical details, such as the type of drainage system.

Regardless, the main contribution of NbS is related to the reduction of the volume of water occurring in the urban sewage system, reducing the risks of local floods (Kaltsidi 2021; Yan et al. 2022), and should be considered during the planning and design phase of urban areas in order to significantly improve their resilience and adaptability to extreme events.

4.4 Stormwater Retention Simulations

Based on the review of tools presented in Sect. 4.2, SWMM has been selected to test different NbS configurations in the urban environment, due to the tool's strengths and its potential for application. In addition, given the high level of detail required by the other software identified and the limited applicability during the design process for designers and urban planners, a native model in Grasshopper has been developed based on physical equations for horizontal NbS (Mosca 2025). Below, the description of the two approaches is given and the application on different tests is illustrated.

(1) **SWMM model:** the Storm Water Management Model (SWMM) was developed in 1971 by the U.S. Environmental Protection Agency (EPA) to help support stormwater management objectives to reduce runoff through infiltration and retention. The model is used for planning, analysis, and design. It allows evaluating grey and green (and hybrid) stormwater control strategies. SWMM allows to predict the performances of various NbS to determine their effectiveness in managing runoff. SWMM can model rain gardens, green roofs, swales, permeable pavements, and other systems, allowing the possibility of adding design data such as soil thickness or porosity (named Low Impact Development, LID, Low Impact Development Center, Inc. 2007; Fletcher et al. 2014). The model allows drawing sub-catchment areas with specific features (e.g. % slope, % impervious area), adding rain gages with time series data, junctions, outfalls and LID control (i.e. NbS or green infrastructure).

(2) **Native Grasshopper approach:** The proposed approach is set up in Grasshopper (McNeel Rhinoceros 3D), and it is based on the assumption that an NbS or green infrastructure can retain a certain amount of stormwater due to its stratigraphy, and can release it to the local plumbing system in a wider time frame compared to a traditional roofing system (Kõiv-Vainik et al. 2022). Among the NbS that can be considered in terms of stormwater retention, particular attention has been given to the ones horizontally oriented and with geometrical features which can naturally collect rainwater, such as green roofs, raingardens, swales, and green areas, due to their geometrical configurations. In addition, these solutions can be implemented at different scales (from the building envelope to open urban areas), and the quantification of their benefits in reducing stormwater runoff should be considered when designing in urban areas (Getter et al. 2007; Longobardi et al. 2019; Zheng et al. 2021). Concerning the construction of the model, a key aspect is the representation of the horizontal NbS, for example, the green roof technology as a 3-dimensional geometry, to allow the calculation of its performance in the Grasshopper parametric interface within McNeel 3D Rhinoceros. To do so, the most relevant features of green roofs need to be identified. According to UNI 11,235 (UNI 2015) (European normative on green roofs), and some of the most relevant green roof producers (in particular, ZinCo, 2020) the most influencing features of green roofs in terms of rainwater collection are:

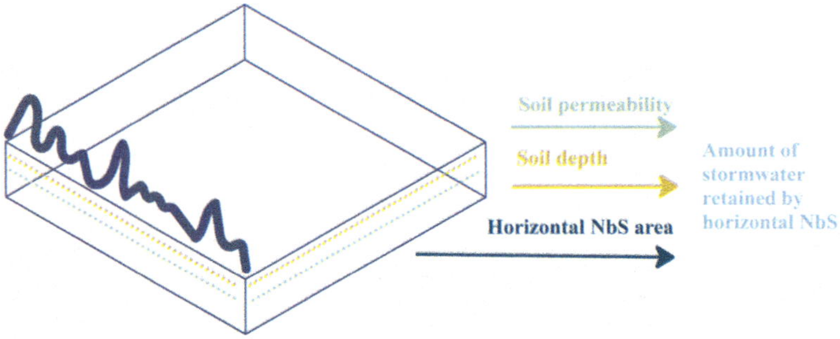

Fig. 4.3 Geometrical representation of green roofs and related most relevant parameters for rainwater collection

- Depth of soil (m);
- Soil storage capacity (m^3/m^3—%) depending on the soil type and related porosity.

In addition, from a performative perspective:

- the surface dedicated to green roofs (m^2) is particularly important to allow, during the decision-making process, to optimise the performance of the proposed design solution (Fig. 4.3).

The model requires technical information about the green roof, such as its water retention capacity based on the technical specifications of the technology used (e.g., Harpo or other types of solutions). It also needs details about the roof's dimensions (including width, length, and depth, depending on the roof type: extensive, semi-intensive, or intensive), as well as information about the layers. Finally, it requires hourly local precipitation data, which is typically readily available from local weather stations. Starting from the information provided as input, the model can calculate the number of litres of stormwater locally retained by the NbS and can calculate the percentage of stormwater which will be released in a wider range of time in the water treatment systems, by reducing the risk of local foods.

More in-depth, the proposed model allows designers to have a real-time evaluation of the quantity of water retained by installing a specific horizontal NbS on a site. Based on the methodology defined by ZinCo green roof systems to assess the performance of their products (ZinCo 2023), the model is based on the following relationships:

$$Q = P \times C \times A \tag{4.1}$$

$$W = A \times h \times C \tag{4.2}$$

where

Q flow rate of the green roof (L/s)
W amount of water retained by the roof (m^3, L)

P precipitation rate (L/s*m^2)
C downflow coefficient (%), defined as = P/D, where P is the precipitation rate (t), and D is the drainage capacity (t). C values vary depending on the technical solution used: for example, for ZinCo solutions, C values can be pre-determined.
A green roof surface (m^2)
H soil depth of the green roof (m).

The methodology for the application and test of the two selected approaches is consistent with the one defined in Chap. 3 for microclimate regulation analysis, in terms of building morphology and the different type of NbS, but also geographical location (typical mediterranean city—Genoa, Italy). The same urban canyon has been considered, with the same orientation and dimensions. The urban canyon is again modelled in the 3D CAD environment using Meshes, for both ground surface and building boxes.

Four configurations of urban canyon with related NbS are modelled and tested: (1) reference canyon (no NbS), (2) green roof and façade (builti-structure NbS), (3) swale (water-media NbS), (4) trees (land-media NbS)—see Fig. 2.1, Chap. 2.

Concerning the input parameters, a typical cumulative precipitation rate has been selected for fall conditions, recognized as the riskiest in terms of extreme rainfalls (Mastrantonas et al. 2021). In addition, the other input data are related to the surface covered by the solution, the depth of the permeable layer and the permeability.

In order to apply the two selected tools to the urban canyon, specific input data have been defined for each (Tables 4.2 and 4.3), and related output data have been collected (Table 4.4). Specifically, the analysis of input and output already provides an overview of the different levels of detail considered by the different approaches.

Results from this application highlight the different level of detail required by the two tested tools to run simulations on NbS (named LID in SWMM), and, subsequently, the different output data provided, and related details. Figure 4.4 shows the graphical visualization of results for the second tool proposed, developed using a colour gradient component in the Grasshopper environment. This feature of the

Table 4.2 Input data for the different canyon configurations (Grasshopper approach)

Configuration	Surface	Precipitation rate	Downflow coefficient	Soil depth
Reference urban canyon	2000 m^2 area	260 mm (mean monthly value for autumn regime)	–	–
Green roof (facade not simulated)	2000 m^2 area, 400 m^2 green roof		0.15	0.3 m
Swale and rain garden	2000 m^2 area, 400 m^2 swale, 200 m^2 raingarden		0.6–0.7	0.3 m
Green areas (grass, trees not simulated)	2000 m^2 area, 200 m^2 green areas		0.3	0.3 m

Table 4.3 Input data for the different canyon configurations (SWMM)

Configuration	Subcatchment area	Impervious surfaces (%)	Slope (%)	LID
Reference urban canyon	2000 m^2	100	0.05	–
Green roof and façade	2000 m^2	80	0.05	Green roof (400 m^2, soil thickness 0.3 m)
Swale and rain garden	2000 m^2	30	0.05	Swale (400 m^2), rain garden (soil thickness 0.3 m)
Green areas (trees not simulated)	2000 m^2	40	0.05	–

Grasshopper approach is particularly relevant and constitute one of the main differences with the other tested tool. The possibility to visualize simulation results can support the selection of the most performing solutions through a visual comparison. Indeed, allowing a real-time visualization of changes in performance of different solutions (type of NbS) or different technical features (e.g. soil depth), depending on input data, can effectively support the decision-making process.

In contrast, the SWMM approach provides data for various objects, such as subcatchments, nodes, links, systems, and variables, including precipitation, snow depth, evaporation, infiltration, runoff, groundwater flow, groundwater elevation, and soil moisture, all on a 15-min timeframe. The most relevant outputs for the urban canyons analyzed in this chapter are presented in Table 4.5. The results show that, for one hour of precipitation (260 mm), runoff and infiltration remain greater than 0 mm for one and a half hours after the precipitation ends. Runoff and infiltration are nearly constant over time (at a 15-min resolution) for canyons 0, 1, and 2. In contrast, in canyon 3, which integrates water media nature-based solutions (including a rain garden and a swale) and only one building, runoff and infiltration behave differently.

Table 4.6 shows the performance of LID systems (referred to as water media nature-based solutions in this book) in terms of total inflow, infiltration loss, surface outflow, initial storage, and final storage, highlighting the different behaviours of these systems (as described in Chap. 2).

Additionally, summary results are provided for various topics, including subcatchment runoff, node depth, node inflow, outfall loading, link flow, flow classification, and conduit surcharge. Within the context of nature-based solutions planning and design, the subcatchment runoff data are particularly interesting (as reported in Table 4.4). The output data include total infiltration, impervious runoff, total runoff, peak runoff, and the runoff coefficient.

Overall, the simulations using the SWMM tool provide a comprehensive range of data, showing that the water management performances of canyons 0, 1, and 2, designed with 100%, 80%, and 40% impervious surfaces, respectively, are similar.

Table 4.4 Output data for the two selected approaches

Configuration	3D Model view	SWMM	GH model
0. Reference urban canyon	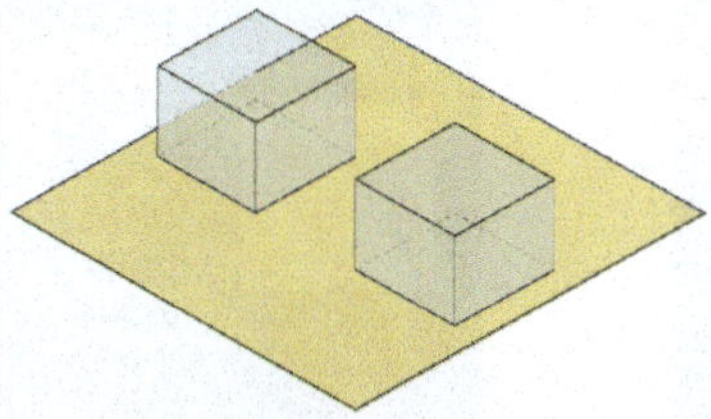	Output: impervious runoff 261.47 mm, pervious runoff 0	Output: 0 m^3 water retained
1. Green roof (green façade not simulated)	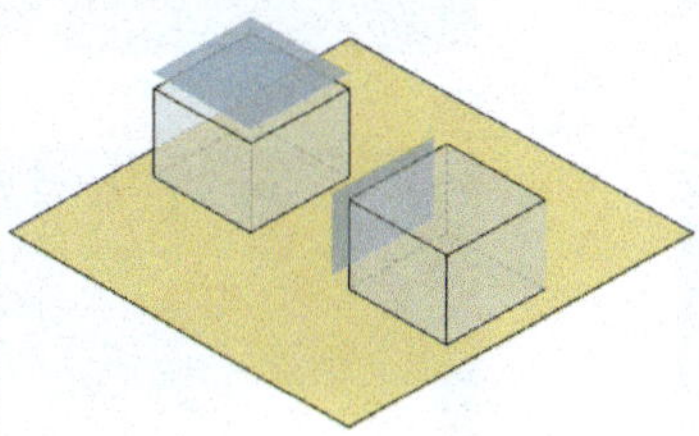	Output: impervious runoff 208.96 mm, pervious runoff 0	Output: 18,120 m^3 water retained (max volume)
2. Swale and rain garden	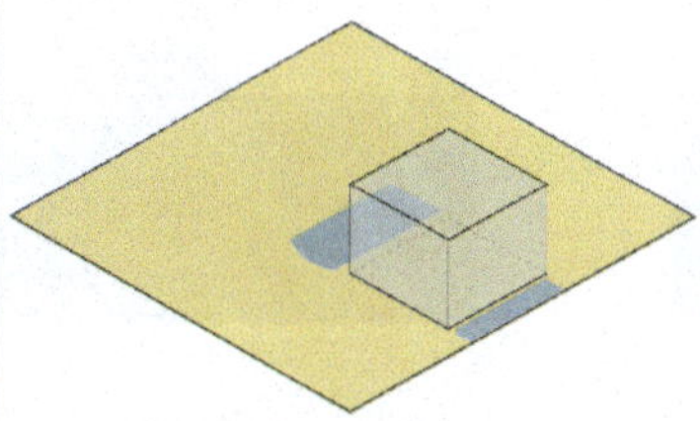	Output: impervious runoff 104.21 mm, pervious runoff 156.36 mm	Output: 34,033 m^3 water retained (max volume for swale + raingarden)
3. Green areas (trees not simulated)	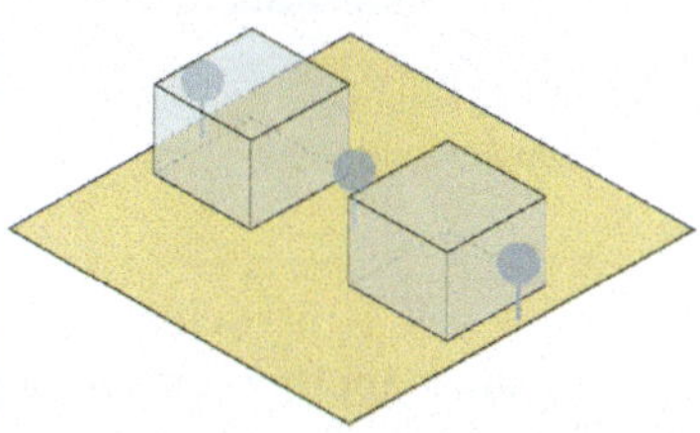	Output: impervious runoff 109.58 mm, pervious runoff 46.83 mm	Output: 12,000 m^3 water retained (max volume)

The modelling of a rain garden and a swale—nature-based solutions specifically designed to manage stormwater—show promising results in canyon 3.

However, a comparison of the results can provide an overview of the performance of the different configurations tested, and supports the quantification of the consistency between the two approaches (Fig. 4.5).

It is worth noting that for the Grasshopper approach, through the use of parametric input data, a specific ratio to define the amount of rainwater potentially collected by a square meter (m^3/m^2) of an NbS with pre-determined technical details can be calculated. Doing so can allow the quantification of the variation in design options performance depending on the soil depth and drainage capacity, streamlining the design process. This aspect can be particularly relevant for the design of green roofs,

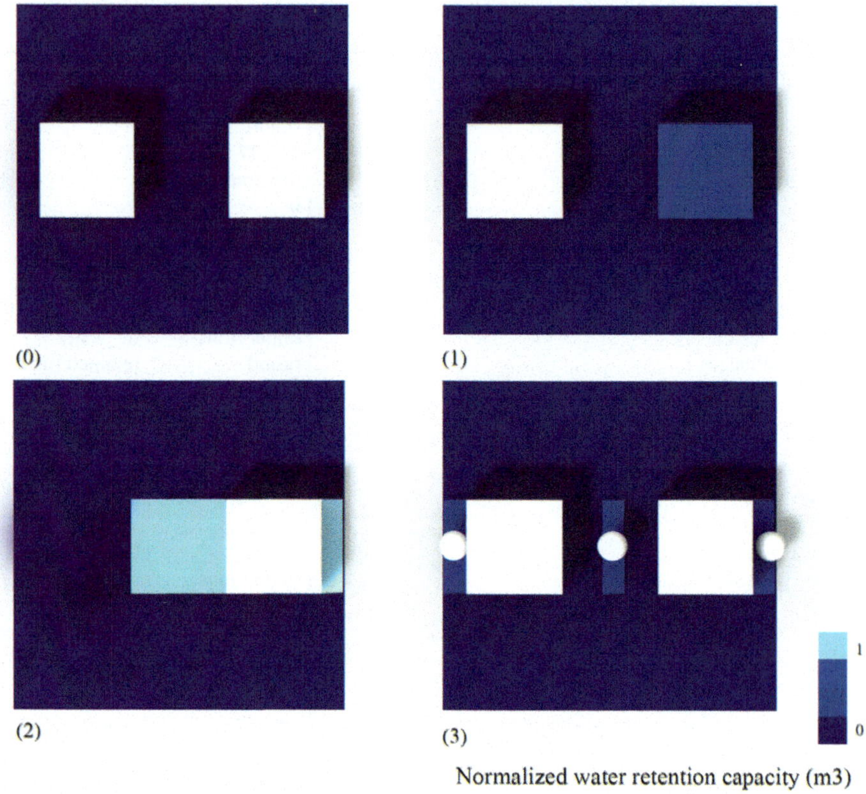

Fig. 4.4 Results visualization provided by the Grasshopper approach (from top left: reference canyon, green roof and façade, swale and raingarden, and trees)

where the optimization of the amount of soil is particularly challenging. During the design process, in order to select the most appropriate solutions, a catalogue of NbS with related "retention ratio" can be produced, and support the decision-making process for maximizing stormwater management on the site.

Among the tested solutions, swales at the ground level are the most efficient in terms of rainwater collection and management, followed by green roofs. Conversely, the introduction of trees does not provide effective performance per se, however, trees can support and optimize the performance of other solutions due to their active captation capacity (Selbig et al. 2022). In addition, the combination of planted trees with permeable pavements or other solutions in the planting sites can improve the performance of this NbS in terms of stormwater management.

The detailed analysis of simulation results highlights some key aspects. In particular, the main result is related to the consistency between the two tested approaches: in both cases, the reference canyon is characterized by the worst conditions, with a total amount of impervious runoff and no infiltration (SWMM), and no rainwater

Table 4.5 Summary of results for runoff (LPS) and infiltration (mm/h) provided by the SWMM approach for a single sub-catchment (four different configuration, see Table 4.4), 15 min timeframe

Elapsed hours	Precipitation (mm/h)	Canyon 0		Canyon 1	
		Runoff (LPS)	Infiltration (mm/hr)	Runoff (LPS)	Infiltration (mm/h)
00:15:00	260.00	144.44	0.00	144.26	0.00
00:30:00	260.00	144.44	0.00	144.30	0.00
00:45:00	260.00	144.43	0.00	144.32	0.00
01:00:00	0.00	144.44	0.00	144.33	0.00
01:15:00	0.00	0.77	0.00	0.46	0.00
01:30:00	0.00	0.16	0.00	0.09	0.00
01:45:00	0.00	0.06	0.00	0.04	0.00
02:00:00	0.00	0.03	0.00	0.02	0.00
02:15:00	0.00	0.02	0.00	0.00	0.00
02:30:00	0.00	0.00	0.00	0.00	0.00
Elapsed hours	Precipitation (mm/h)	Canyon 2		Canyon 3	
		Runoff (LPS)	Infiltration (mm/h)	Runoff (LPS)	Infiltration (mm/h)
00:15:00	260.00	143.35	0.77	314.27	0.66
00:30:00	260.00	144.13	0.47	110.00	0.41
00:45:00	260.00	144.22	0.36	122.09	0.31
01:00:00	0.00	144.26	0.32	110.12	0.28
01:15:00	0.00	3.96	0.31	2.79	0.27
01:30:00	0.00	0.95	0.30	0.90	0.26
01:45:00	0.00	0.33	0.30	0.19	0.26
02:00:00	0.00	0.12	0.30	0.06	0.26
02:15:00	0.00	0.04	0.30	0.00	0.26
02:30:00	0.00	0.00	0.30	0.00	0.00

Table 4.6 Focus on LID performances for rain garden and swale in the third configuration of canyon

LID control	Total inflow mm	Infiltration loss mm	Surface outflow mm	Initial storage mm	Final storage mm
Raingarden	306.83	0.00	305.25	30.00	31.58
Swale	306.83	0.89	623.21	0.00	0.03

retained (GH), conversely, swales and raingardens are the most performing solutions, in terms of ratio between impervious and pervious runoff, infiltration and amount of water retained by these systems. Finally, green roofs provide a limited performance

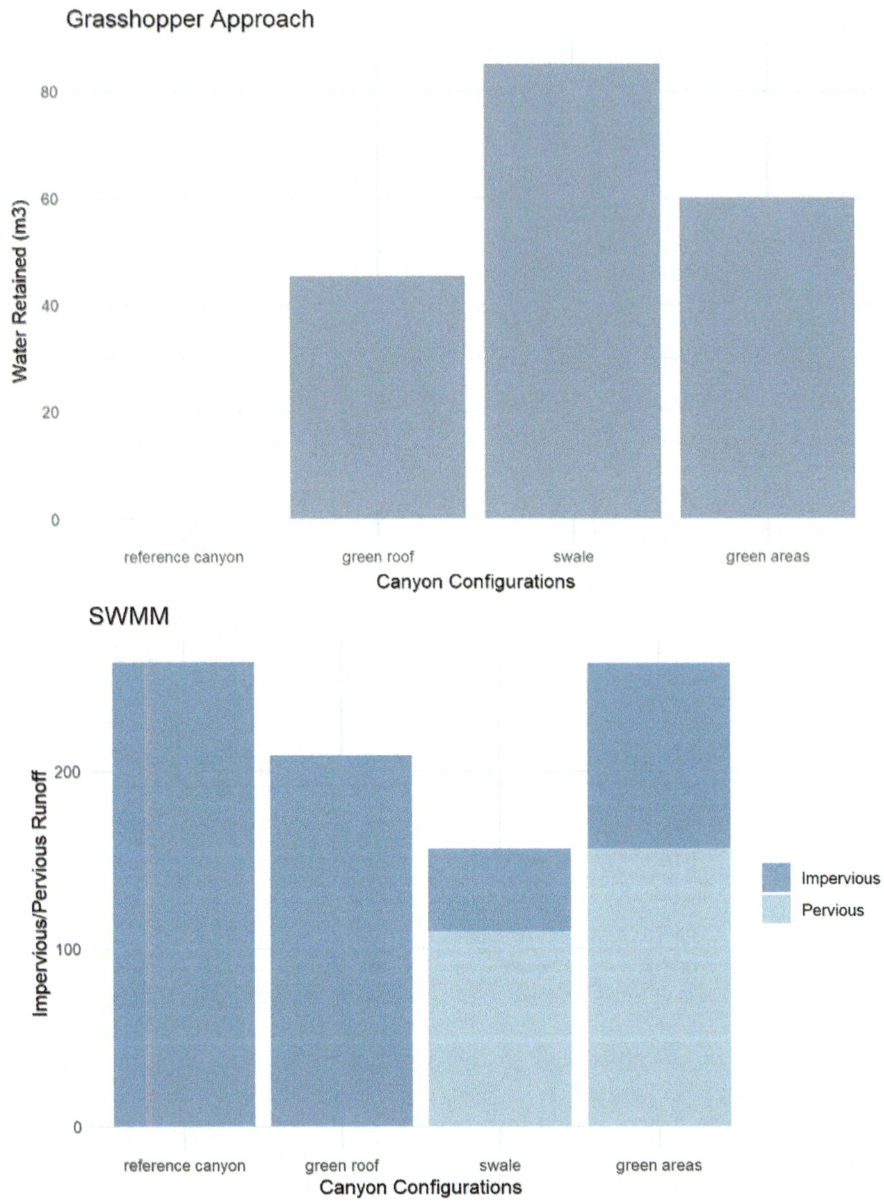

Fig. 4.5 Bar graph with comparison of the four canyon configuration tested with the two approaches (Grasshopper and SWMM, elaborated with RStudio with the ggplot2 function—Rstudio, 2024, Wickham 2016)

compared to other water-based NbS, however, in very dense urban areas can effectively supports the reduction of surface runoff through their capacity to temporarily store rainwater (Longobardi et al. 2019).

From the perspective of the accuracy of results, it is worth mentioning that the SWMM model has been validated by different researchers though field measurement and calibration activities (Kourtis et al. 2017; Zakizadeh et al. 2022). Therefore, given the consistency between the two models (as shown in Fig. 4.3), also the other proposed approach can be considered acceptable for the quantification (prediction) of a design solution at early design stages, especially when the level of detail is still limited.

Finally, a comparison of the two approaches is carried out in relation to target objectives. The main goal of this final consideration is to evaluate the effectiveness of these tools for specific design stages (namely early design stages and detailed design), and identify the trade-offs between the two approaches in terms of different aspects. The key criteria identified for this comparison are first related to the computational power needed, then, other relevant aspects such as the usability of the tools, the level of detail needed and provided and, finally, the background knowledge needed for the use of the selected tools are identified (Table 4.7).

The results shown in Table 4.5, derived from the testing of the two approaches on different NbS options, highlight some relevant aspects for drawing conclusions in this field. In particular, given the ease of use of both approaches in terms of computational requirements, it could be proposed to use each of the two approaches for different situations and needs. For example, during the preliminary design phase (in terms of built volumes, NbS types, and their extent), the approach in Grasshopper can provide valuable feedback to the designer for optimizing the layout of spaces and solutions on urban land. Similarly, once the preliminary design is defined and the level of detail is increased (for example, materials used, stratigraphy), a more comprehensive approach like SWMM can be implemented to further explore the performance of different solutions, not only in terms of water retention but also surface runoff. SWMM also allows for evaluating water retention performance at the district scale,

Table 4.7 Comparison of the two described and tested approaches, in relation to the criteria selected

Criteria	GH approach	SWMM model
Computational power required	Low	Low
Usability in terms of design complexity	High	Low
Input data (amount of data required)	Low	Medium–high
Output data (type and accuracy)	Q = flow rate for an NbS (L/s), W = amount of water retained by the roof (m3)	Runoff (LPS), groundwater Flow (LPS), Groundwater Elev (mm), etc
Level of knowledge required	Low	Medium
Type of background	Design	Engineering

with varying levels of detail, including in terms of hydraulic components, particularly in high-risk areas for stormwater management, thanks to the involvement of relevant expertise.

4.5 Summary

This chapter provides a comprehensive overview of the challenges in urban areas related to stormwater management, and the important role of NbS in stormwater retention and runoff regulation. Through a literature review, the most relevant parameters and tools for evaluating the performance of water-based NbS have been identified and described. A series of applications on various urban canyon configurations with different types of NbS is provided to highlight the trade-offs and synergies between the two approaches and the different NbS tested. The main potentials and limitations of the selected simulation tools for the calculation of stormwater retention are identified and discussed, and recommendations for their application during the design process are provided. Finally, an analysis of the performances of the different NbS tested is provided, identifying the most performing solutions.

References

Allan RP, Barlow M, Byrne MP et al (2020) Advances in understanding large-scale responses of the water cycle to climate change. Annu NY Acad Sci 1472:49–75. https://doi.org/10.1111/nyas.14337

Biswal BK, Bolan N, Zhu Y-G, Balasubramanian R (2022) Nature-based Systems (NbS) for mitigation of stormwater and air pollution in urban areas: a review. Resour Conserv Recycl 186:106578. https://doi.org/10.1016/j.resconrec.2022.106578

CNT (2021) Green values calculator. In: Center for neighborhood technology. https://cnt.org/tools/green-values-calculator. Accessed 27 Jan 2025

Dotto CBS, Kleidorfer M, Deletic A et al (2011) Performance and sensitivity analysis of stormwater models using a Bayesian approach and long-term high resolution data. Environ Model Softw 26:1225–1239. https://doi.org/10.1016/j.envsoft.2011.03.013

Douville H, Raghavan K, Renwick J et al (2021) Water cycle changes. In: Masson-Delmotte VP, Zhai P, Pirani A et al (eds) Cambridge University Press, Cambridge, pp 1055–1210

Engel T (2016) Long term hydrologic impac analysis. https://engineering.purdue.edu/~lthia/. Accessed 27 Jan 2025

Fletcher TD, Shuster W, Hunt WF, et al (2014) SUDS, LID, BMPs, WSUD and more—the evolution and application of terminology surrounding urban drainage. Urban Water J 1–18

Fletcher TD, Shuster W, Hunt WF et al (2015) SUDS, LID, BMPs, WSUD and more—the evolution and application of terminology surrounding urban drainage. Urban Water J 12:525–542. https://doi.org/10.1080/1573062X.2014.916314

Getter KL, Rowe DB, Andresen JA (2007) Quantifying the effect of slope on extensive green roof stormwater retention. Ecol Eng 31:225–231. https://doi.org/10.1016/j.ecoleng.2007.06.004

Hoerling M, Barsugli J, Livneh B et al (2019) Causes for the century-long decline in Colorado river flow. J Clim 32:8181–8203. https://doi.org/10.1175/JCLI-D-19-0207.1

IPCC (2021) Climate change 2021. The Physical Science Basis. https://www.ipcc.ch/report/ar6/wg1/downloads/report/IPCC_AR6_WGI_Full_Report.pdf. Accessed 27 Oct 2021

Jayasooriya VM, Ng AWM (2014) Tools for modeling of stormwater management and economics of green infrastructure practices: a review. Water Air Soil Pollut 225:2055. https://doi.org/10.1007/s11270-014-2055-1

Kaltsidi MP (2021) Assessment and optimization of felt living walls in terms of water retention performance and artificial lighting

Kõiv-Vainik M, Kill K, Espenberg M et al (2022) Urban stormwater retention capacity of nature-based solutions at different climatic conditions. Nature Based Solut 2:100038. https://doi.org/10.1016/j.nbsj.2022.100038

Koop SHA, van Leeuwen CJ (2017) The challenges of water, waste and climate change in cities. Environ Dev Sustain 19:385–418. https://doi.org/10.1007/s10668-016-9760-4

Kourtis IM, Kopsiaftis G, Kopsiaftis G et al (2017) Calibration and validation of SWMM model in two urban catchments in Athens, Greece. In: 15th International conference on environmental science and technology. https://www.researchgate.net/profile/Ioannis-Kourtis/publication/319465255_Calibration_and_validation_of_SWMM_model_in_two_urban_catchments_in_Athens_Greece/links/59ad33ef0f7e9bdd115c264f/Calibration-and-validation-of-SWMM-model-in-two-urban-catchments-in-Athens-Greece.pdf. Accessed 12 Mar 2025

Krull W, Berry P, Bauduceau N et al (2015) Towards an EU research and innovation policy agenda for nature-based solutions & re-naturing cities. Final report of the Horizon 2020 expert group on nature-based solutions and re-naturing cities. LUX

Longobardi A, D'Ambrosio R, Mobilia M (2019) Predicting stormwater retention capacity of green roofs: an experimental study of the roles of climate, substrate soil moisture, and drainage layer properties. Sustainability 11:6956. https://doi.org/10.3390/su11246956

Low Impact Development Center, Inc. (2007) LID urban design tools—background. http://www.lid-stormwater.net/background.htm. Accessed 7 Jun 2016

Madsen H, Lawrence D, Lang M et al (2014) Review of trend analysis and climate change projections of extreme precipitation and floods in Europe. J Hydrol 519:3634–3650. https://doi.org/10.1016/j.jhydrol.2014.11.003

Maltby E, Barker T (2009) The Wetlands handbook, 2 volume set. Wiley

Mannina G, Viviani G (2010) An urban drainage stormwater quality model: model development and uncertainty quantification. J Hydrol 381:248–265. https://doi.org/10.1016/j.jhydrol.2009.11.047

Massoudieh A, Nguyen K, Murthy S (2023) An extensible, plugin-based tool for modeling flow and reactive transport in water systems. Environ Model Softw 164:105707. https://doi.org/10.1016/j.envsoft.2023.105707

Mastrantonas N, Herrera-Lormendez P, Magnusson L, Pappenberger F, Matschullat J (2021) Extreme precipitation events in the Mediterranean: spatiotemporal characteristics and connection to large-scale atmospheric flow patterns. Int J Climatol 41(4):2710–2728. https://doi.org/10.1002/joc.6985

Mosca F (2025) Innovative approaches for the environmental assessment of ecological building envelopes—a computational application in the ECOLOPES project. PhD dissertation

Novák V, Hlaváčiková H (2019) Soil-water content and its measurement. In: Applied soil hydrology. Springer International Publishing, Cham, pp 49–61

Oppenheimer J, Badruzzaman M, McGuckin R, Jacangelo JG (2014) Urban water-cycle energy use and greenhouse gas emissions. J AWWA 106:E86–E97. https://doi.org/10.5942/jawwa.2014.106.0017

Palermo SA, Turco M, Pirouz B et al (2023) Nature-based solutions for urban stormwater management: an overview. IOP Conf Ser Earth Environ Sci 1196:012027. https://doi.org/10.1088/1755-1315/1196/1/012027

Pitt R, Voorhees J (2002) SLAMM, the source loading and management model. Wet-weather flow in the urban watershed, pp 103–139

Posit PBC (2024) RStudio: Integrated development environment for R (Version 2024.x). https://posit.co
Prosdocimi I, Kjeldsen TR, Miller JD (2015) Detection and attribution of urbanization effect on flood extremes using nonstationary flood-frequency models. Water Resour Res 51:4244–4262. https://doi.org/10.1002/2015WR017065
Qin Y (2020) Urban flooding mitigation techniques: a systematic review and future studies. Water 12:3579. https://doi.org/10.3390/w12123579
Razguliaev N, Flanagan K, Muthanna T, Viklander M (2024) Urban stormwater quality: a review of methods for continuous field monitoring. Water Res 249:120929. https://doi.org/10.1016/j.watres.2023.120929
Ridolfi E (2014) Exploring the urban hydrosocial cycle in tourist environments. Ingeo 17. https://doi.org/10.14198/INGEO2014.61.02
Selbig WR, Loheide SP, Shuster W et al (2022) Quantifying the stormwater runoff volume reduction benefits of urban street tree canopy. Sci Total Environ 806:151296. https://doi.org/10.1016/j.scitotenv.2021.151296
Šimůnek J, van Genuchten M Th, Šejna M (2008) Development and applications of the HYDRUS and STANMOD software packages and related codes. Vadose Zone J 7(2):587–600. https://doi.org/10.2136/vzj2007.0077
UNEP (2018) Nature-based solutions for water management
UNI (2015) UNI 11235: criteria for design, execution, testing and maintenance of roof gardens. Ente Italiano di Normazione (UNI)
US EPA O (2014) Storm water management model (SWMM). https://www.epa.gov/water-research/storm-water-management-model-swmm. Accessed 9 Sep 2024
US EPA O (2015) What is green infrastructure? https://www.epa.gov/green-infrastructure/what-green-infrastructure. Accessed 24 Oct 2023
Versini P-A, Gires A, Tchinguirinskaia I, Schertzer D (2016) Toward an operational tool to simulate green roof hydrological impact at the basin scale: a new version of the distributed rainfall–runoff model Multi-Hydro. Water Sci Technol 74:1845–1854. https://doi.org/10.2166/wst.2016.310
Versini P-A, Kotelnikova N, Poulhes A et al (2018) A distributed modelling approach to assess the use of Blue and Green Infrastructures to fulfil stormwater management requirements. Landsc Urban Plan 173:60–63. https://doi.org/10.1016/j.landurbplan.2018.02.001
Vijayaraghavan K, Biswal BK, Adam MG et al (2021) Bioretention systems for stormwater management: recent advances and future prospects. J Environ Manage 292:112766. https://doi.org/10.1016/j.jenvman.2021.112766
Wickham H (2016) Ggplot2. Springer International Publishing. https://doi.org/10.1007/978-3-319-24277-4
Yan J, Zhang S, Zhang J et al (2022) Stormwater retention performance of green roofs with various configurations in different climatic zones. J Environ Manage 319:115447. https://doi.org/10.1016/j.jenvman.2022.115447
Zakizadeh F, Moghaddam Nia A, Salajegheh A et al (2022) Efficient urban runoff quantity and quality modelling using SWMM model and field data in an urban watershed of Tehran metropolis. Sustainability 14:1086. https://doi.org/10.3390/su14031086
Zhang S, Lin Z, Zhang S, Ge D (2021) Stormwater retention and detention performance of green roofs with different substrates: Observational data and hydrological simulations. J Environ Manage 291:112682. https://doi.org/10.1016/j.jenvman.2021.112682
Zheng X, Zou Y, Lounsbury AW et al (2021) Green roofs for stormwater runoff retention: a global quantitative synthesis of the performance. Resour Conserv Recycl 170:105577. https://doi.org/10.1016/j.resconrec.2021.105577
ZinCo (2023) Planning guide, water retention with a "Stormwater Management Roof". www.zinco-greenroof.com
ZinCo (2020) Green roof solutions: sustainable urban development. ZinCo GmbH

Chapter 5
Performance-Driven Design Approach: A Case Study

Abstract This chapter introduces a workflow for a performance-driven design process, considering both microclimate regulation and human comfort (Chap. 3) and stormwater retention (Chap. 4) as key environmental parameters for climate adaptation. An urban case study is included to better explain the approach step by step, present research results, and derive design recommendations, including best-performing design strategies. The presented approach demonstrates how nature-bases solutions can be selected and designed based on measurable Key Performance Indicators, offering a practical and applied perspective. Design considerations are provided starting from the application of the workflow on the selected case study, and related conclusions are drafted.

Keywords Nature-based solutions · Environmental analysis · Climate adaptation · Environmental design · Microclimate · Stormwater management

5.1 Desing Process and Computational Workflow

The design of nature-based cities for climate change adaptation and environmental risk management and prevention can be informed and supported, as described in Chaps. 3 and 4, by computational approaches. The design approach can focus on different Key Performance Indicators (Chap. 2), with site specific thresholds. Balancing the design choices with the several input provided by simulations outcomes, client requirements, site specific constraints, etc., will be among the most difficult challenge. As it will be shown, besides the type, surface extension, and location of nature-based solutions, also urban and building morphological features can influence environmental performance, especially in the case of microclimatic conditions (Mosca et al. 2025).

A performance-driven design approach, schematically presented in Fig. 5.1, involves first the collection of site-specific data, such as climatic and precipitation data. This data can include average values for specific seasons (e.g., summer or winter), annual data, or data from extreme weather events. The latter is particularly

K. Perini and F. Mosca, *Nature-Based Cities*, Digital Innovations in Architecture, Engineering and Construction, https://doi.org/10.1007/978-3-031-94612-7_5

relevant for managing environmental risks associated with extreme weather, which is the second most significant global risk over a two-year period and the primary risk over a ten-year period (Global Economic Forum 2024). Examples of extreme weather include heatwaves and heavy precipitation (Intergovernmental Panel On Climate Change (IPCC) 2023). Selecting a particularly hot day for site analysis simulations can yield valuable insights into the microclimatic conditions during a heatwave. Additionally, computationally processed data, such as climate projections (Eyring et al. 2016) can be utilized to predict future scenarios, supporting the implementation of adaptable solutions. Simulating microclimatic conditions under varying circumstances (e.g., typical winter or summer days, or the hottest summer day) can provide essential data for designing climate change adaptation strategies. In the case of water management related simulations, the data collection can include average rain fall/month/day/hour and also extreme events with high return periods (e.g. 100 years up to 10,000 years). The return period is also known as a recurrence interval and it is an estimate of the likelihood of an event, such as storm, to occur (Ma et al. 2019) and allows to identify storm-induced hotspot (Ferreira et al. 2018).

The second step of the described design process (Fig. 5.1) involves the site analysis in terms of urban morphology and land use. The buildings surrounding a site can create shadows and alter wind patterns, significantly affecting microclimatic conditions (Eyring et al. 2016). Land use not only influences Urban Heat Island (UHI) phenomena (Taha 1997; Dale et al. 2025), but also impacts water management (Cao et al. 2024). The use of waterproofing materials common in urban areas prevents water from evaporating or infiltrating into groundwater, leading to increased runoff (Federal Interagency Stream Restoration Working Group 1998). Site analysis data

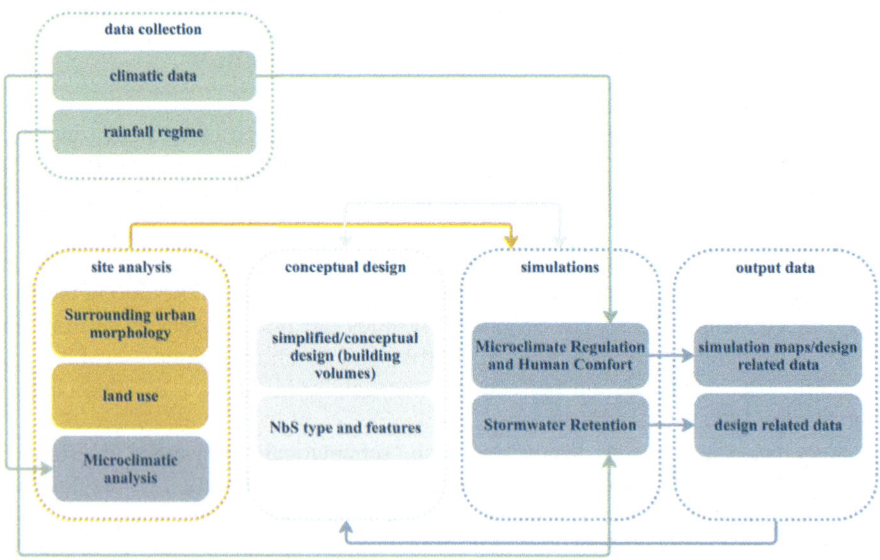

Fig. 5.1 Simplified diagram of performance-driven design approach

can be interpreted thoroughly by means of microclimatic simulations, using the same approach described in the simulations step of this process and deeply described in Sect. 4.3.

The third step aims at developing a simplified or conceptual design to determine building volumes and uses. This step will include all the client requirements and municipalities constrains, according to the design brief, for instance, by defining the necessary floor space, functions, etc. (Canepa et al. 2022). In this step, the types of nature-based solutions (NbS, e.g., green roofs, trees, rain gardens) will also be defined and strategically located, in alignment with the outdoor space and building uses (e.g., a public park surrounding a residential building). The design and selection of NbS should be verified and, if needed, re-designed for optimization, by means of the fourth step simulations.

The simulation step is thoroughly outlined and represented by the computational workflow (Fig. 5.2). This book primarily focuses on microclimate regulation for human comfort and on stormwater retention. However, this approach can include other Key Performance Indicators, such as carbon sequestration (Baggaley et al. 2022), or air (Barwise and Kumar 2020) and water pollution (Sharma and Malaviya 2021) reduction. It is worth noting that the two simulation approaches presented—microclimate regulation and human comfort, and stormwater retention—address adaptation to the climate change drivers and effects increased temperature and altered precipitation (Directorate-General for Research and Innovation (European Commission) et al. 2021). These approaches contribute to managing extreme weather risks impacting the built environment. The inclusion of other KPIs related to additional climate change effects and drivers, such as carbon storage and sequestration in vegetation for GHC concentration, can help prevent risks arising from anthropogenic activities.

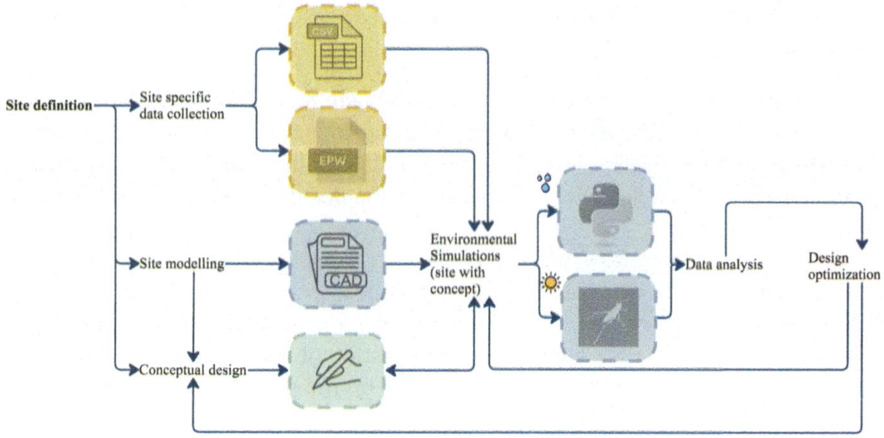

Fig. 5.2 Computational workflow for performance-driven design approach

The data obtained through the simulations can be represented by heatmaps and by site specific data, such as air temperature (°C), UTCI (°C), stormwater retained (L), as described in Chaps. 3 and 4. These data are highly relevant to inform the design process. For example, several design configurations can be modelled and simulated for comparison, as shown in Chap. 4, Sect. 4.3. The presented approach can be applied to both new construction projects at various scales (from district/neighbourhood to building level) and refurbishment design projects.

Based on the approaches defined in Chaps. 3 and 4 and on the design approach outlined in Fig. 5.1, a computational workflow has been developed to apply environmental analysis in a climate change adaptation perspective, both in relation to microclimate regulation and stormwater management (Fig. 5.2).

Starting from the definition of the site, specific environmental data can be collected through local weather stations (provided as .csv) or from weather files (provided as.epw), relevant for the current state analysis and simulation. In parallel, through the use of a CAD interface, the urban morphology of the selected site is modelled (0.3 dm), and a first draft of conceptual design is usually manually defined.

Based on the information collected, environmental data are used to simulate the microclimatic and water management conditions of the site. Output data, provided as .csv data, can be analysed and used to optimize the design solution, resulting in an iterative process based on loops between environmental simulations and the design development.

From an operational perspective, the necessary data to apply this performance quantification process for design solutions are as follows:

- **For calculating the outdoor comfort index:** geolocation, climatic data of typical or extreme summer day, and urban morphology. Additionally, the geometry of the proposed building and the potential presence of NbS.
- **For calculating stormwater retention**: typical or extreme rainfall regime, type of NbS suitable for reducing surface runoff, surface area, soil depth, and the drainage capacity of the implemented system.

The outputs provided by the described workflow are related to the shading capacity of the solution as well as the amount of rainwater retained by the different solutions implemented at the different scales. Output data can be used to inform the design process and guide the designers in optimising the design to achieve local resilience needs (Fig. 5.2).

The presented workflow highlights the importance of computational tools to quantify NbS performance at early design stages, in order to achieve an optimal balance between microclimatic and stormwater retention improvement capacity (or other environmental KPIs). Computationally, different approaches and tools can be further implemented in order to automatize the optimization process, depending on specific design needs. For example, multi-objective optimization can be implemented in order to give equal priority to the different environmental objectives and reduce the trade-offs of NbS performance in relation to the different aspects considered (Chang 2015; Weerasuriya et al. 2021).

On the other hand, a site can have specific requirements and needs, also from an environmental perspective: for example, depending on the urban morphology and other contextual conditions, designers could need to prioritize NbS performance in terms of microclimate regulation for heat-related stress reduction, or stormwater management and runoff reduction. In this context, a multi-attribute optimization can be implemented in order to prioritize design solutions depending on specific objectives and requirements (Kazimieras Zavadskas et al. 2019; Vinogradova 2019; Hwang and Yoon 1981a, b).

5.2 Case Study Application

In order to apply and showcase step by step the presented performance driven design approach and related computational workflow, a site has been selected, and three design options developed. The case study is located in Genoa, Italy, a city of medium scale characterized by a temperate mediterranean climate, with warm summers and mild winters, and precipitations condensed especially during the autumn season.

5.2.1 Data Collection

Defining geographical and climatic conditions of the case study, in addition to the urban morphology of the area, is crucial when a design proposal aims at improving the capacity to adapt to local effects of climate change. The Genoa Metropolitan Area has been historically affected by floods due to two main factors: meteorological conditions and the city's geomorphology. The meteorological cause is a secondary cyclone that forms or intensifies from an existing cyclone south of the Alps, influenced by orographic effects. Geographically, the city's narrow coastal area is bordered by steep mountains, which further exacerbate flooding (Acquaotta et al. 2018).

The data collection primarily includes climatic data of the site. Depending on the location of the weather stations, these data may vary in relevance, as they can be influenced by the site's specific morphological and environmental conditions. For the case study presented, it was possible to collect data from a weather station nearby the design site. Climatic data for microclimatic simulations include: air temperature (hourly), relative humidity (hourly) and wind speed and direction (average daily). Figure 5.3 shows the diagrams related to a very hot summer day, specifically 11th August 2024. The precipitation data collected for the case study come from the same weather station considered for the climatic data; in particular, monthly accumulated rainfalls for 2024 have been collected, and data from October (typical autumn conditions, where precipitation are usually more intense) have been selected (260.8 mm).

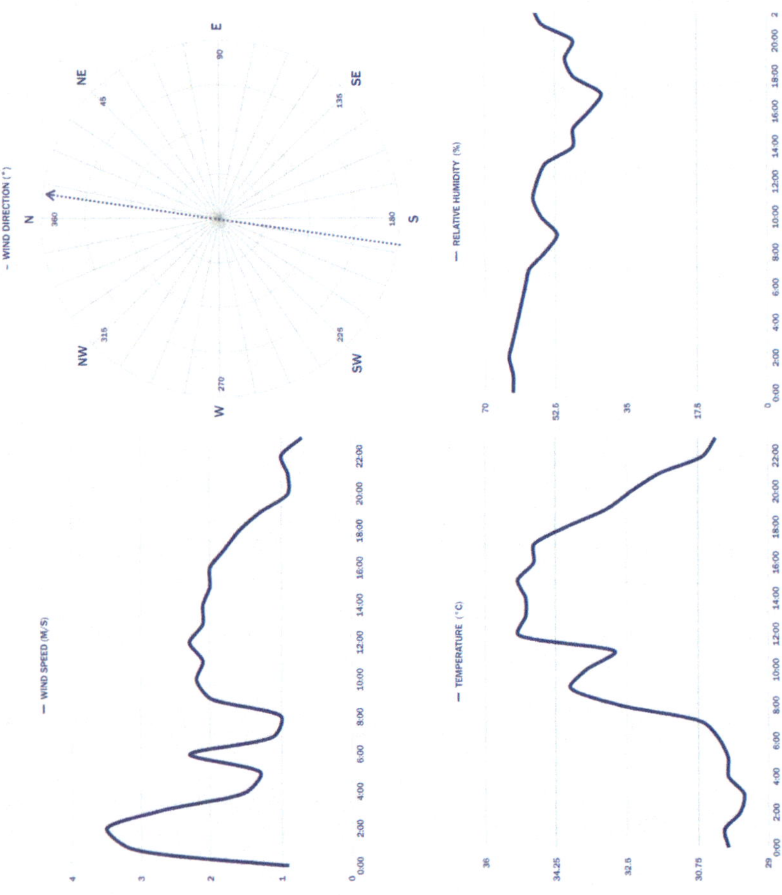

Fig. 5.3 Climatic data collected for the case study, Genoa city center, 11th August 2024 (Credit: UNIGE students C. Galleano, E. Macharashvili, L. Golinja, S.-F. Rashidi)

5.2.2 Site Analysis

Within the area of the Municipality, a district in the city centre has been selected due to the regular urban morphology as well as the orography, with dense and compact midrise buildings, distributed following a conventional grid layout and a limited amount of green spaces (Selvan et al. 2024). The area is located near the seaside and is primarily residential, with buildings ranging from six to eight stories in height. Urban canyons with single or double tree lines are present. However, the area lacks greenery near common spaces and is quite dense, with roads that are accessible by cars. It is worth noticing that the area is characterized by a medium level of risk for floodings, given the proximity with one of the major urban creek of the city and the sea. From a normative perspective, the selected area falls under a transformation regime. Specifically, a former bus depot of the urban public transport company, now disused, is planned for demolition to make way for a new mixed-use residential development. The regulatory constraints allow for a maximum site coverage of 50%, while the height of the new building must not exceed that of the surrounding structures.

Additionally, the site's classification as a medium-risk flood zone necessitates the implementation of design solutions aimed at enhancing liveability and mitigating associated risks (Fig. 5.4).

An analysis of the microclimatic conditions shows the percentage of shaded area on the site as well as the distribution of shadows provided by existing buildings on open areas in different seasons (Fig. 5.5). The result of this analysis shows that most of the area is not shadowed by the surrounding buildings, due to its dimensions and the height of the buildings, during summer.

Finally, from an ecological perspective, a preliminary analysis of existing species on the site has been conducted. The main aim of this activity is to collect information on existing flora and fauna, and derive the most appropriate plant species to be included in the design. This information is particularly relevant for the detailed modelling and simulation of NbS performance, for microclimate regulation but also for stormwater retention. Indeed, depending on the plant species selected, the thermal performance can vary, similarly, for a detailed design of NbS for stormwater regulation and rainwater collection, the selection of adequate species is crucial. In order to collect these information, the Global Biodiversity Information Facility (GBIF, GBIF.org 2025) database is used.

By selecting the area of interest through the use of geographical coordinates, a list of occurrences for different taxonomic groups can be found. Results from this preliminary analysis demonstrated that a wide range of species can be found on the site and the surrounding area, among plants, animals, fungus, and other smaller organisms (Table 5.1). Specifically, a focus on the plant species has been conducted for this application, however, other living organisms such as mammals, insects and micro-organisms can be further analysed and included in the design process when biodiversity-related indicators are considered (Table 5.1).

It is worth mentioning that this analysis is also relevant in order to identify native species and reduce risks of including alien species, which can have a negative impact

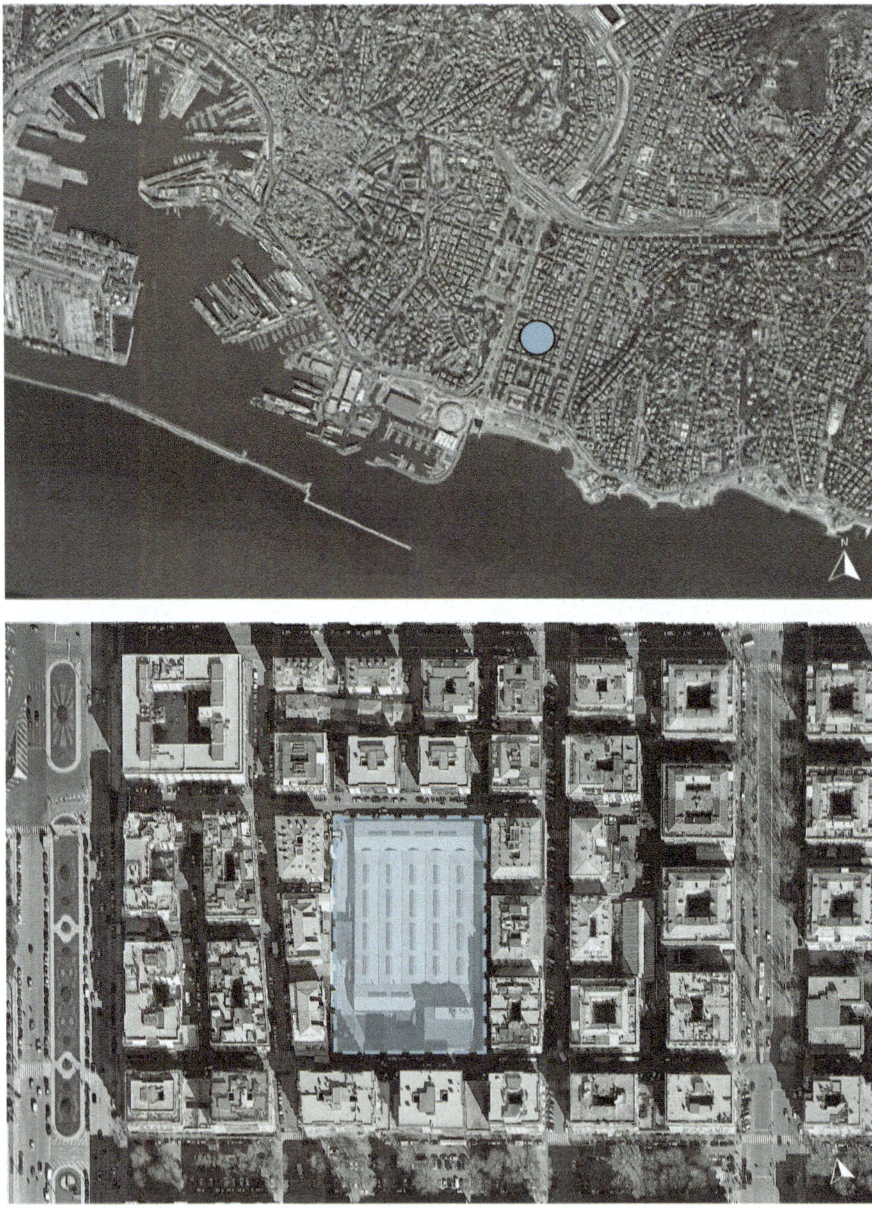

Fig. 5.4 Case study site top view, Genoa (Italy)

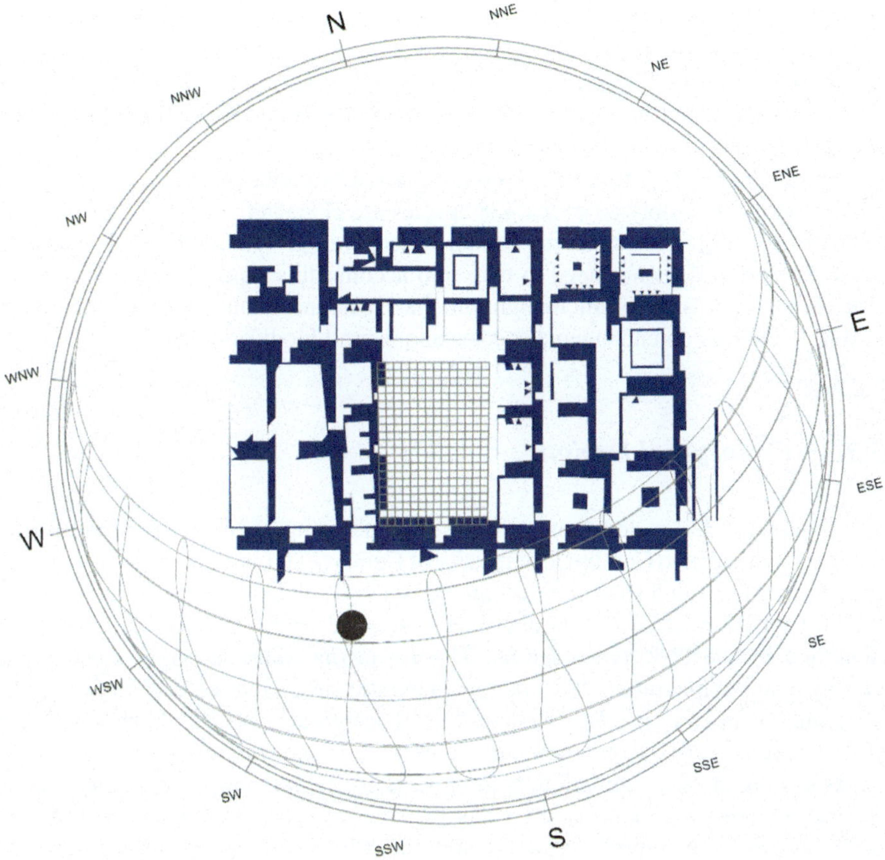

Fig. 5.5 Current state shadow analysis performed with Ladybug Tools (Rhinoceros 3D), for the 21st June, 2024, 08:00

Table 5.1 Overview of taxonomic distribution of occurrences on the site (GBIF.Org User 2025)

Taxonomic group	Major groups	Occurrences
Plants	Tracheophyta (421)	428
	Chlorophyta (2)	
	Rhodophyta (2)	
	Charophyta (2)	
	Bryophyta (1)	
Animals	–	156
Fungus	–	18

on local ecosystems. Specifically, for this case study, particular attention has been given to existing tree species, given the necessity to model them for the assessment of microclimatic conditions of the site. Species such as *Olea europea*, *Quercus ilex*, *Acacia dealbata*, and *Prunus domestica* can be found, and constitute a starting point for the selection of plant species for the design.

Among the plants listed, an analysis of the IUCN Global Red List Category can also be performed: only 83 of the 428 species are characterized by a least concern, while no existing species are considered endangered or vulnerable. In this specific case, the design does not need to take into account this aspect, given the reduced risks, however, it is important to highlight that a designer should consider the impact of design choices on endangered species and overall biodiversity loss.

5.2.3 Conceptual Design

The conceptual design development is based on planning and design criteria outlined the Piano Urbanistico Comunale of the city of Genoa (PUC, issued in 2000). Table 5.2 presents the identified values, which are considered solely for the purpose of testing the performance-driven design approach discussed in this chapter. The main use identified by the PUC is residential. The 1.4 m^2/m^2 value indicates the building density, and the maximum 50% lot coverage ratio shows that at least half of the lot area must be open space. Other planning and design criteria include the maximum usable floor area, building height, and minimum distance requirements.

According to the regulation requirements, different design solutions can be defined, meeting the mentioned criteria. In order to compare the design solutions, different typologies/morphologies are adopted (Fig. 5.6).

(1) Terraced building;
(2) Two-blocks;
(3) Three-blocks.

Table 5.2 Planning and design criteria (based on Municipal Plan of Genoa, namely "Piano Urbanistico Comunale" (PUC)

Design criteria	Value
Settlement Utilization Index	1.4 m^2/m^2
Usable floor area	11,606 m^2
Distances	1.5 m from property boundaries 5.0 m from vehicular roads
Maximum coverage ratio	The coverage ratio must not exceed 50% of the lot
Heights	The heights of the new building must not exceed those of the surrounding buildings
Minimum internal height	Not less than 2.7 m for habitable rooms, reduced to 2.4 m for direct accessory spaces

The design of each solution was based on key objectives, including improving human-nature interactions, optimizing indoor daylight, enhancing water management, ensuring thermal comfort during the summer, and increase biodiversity. The objective of optimizing indoor daylight was achieved by orienting the building facades and most of the windows toward the south, with a high window-to-wall ratio.

The goal of improving human-nature interactions was primarily achieved in the first and third design solutions by incorporating accessible terraces with green roofs. The remaining key objectives—enhancing water management and ensuring thermal comfort during the summer—were addressed through the integration of nature-based

Fig. 5.6 Three different building designs/morphologies developed for the selected site according to the design requirements (credits for 3D design: UNIGE students C. Galleano, E. Macharashvili, L. Golinja, S. Rashidi, A. Hassanein, A. Awawdeh, G. Ghita, R. Alleithy, G. Caprile, A. Sogand, M. Squeo)

Fig. 5.6 (continued)

Fig. 5.6 (continued)

solutions, primarily through landscaping and built structures, to reduce impermeable surfaces.

In the case of the first design solution, a terraced building could be designed with a uniform height and depth to enclose the site while maximizing floor area. The courtyard can constitute an opportunity to include green space, communal areas, or parking. In addition, a mix of uses (mainly residential, with commercial, or office) can be included along the outer façade to interact with the surrounding urban fabric, offering flexibility in unit layouts, allowing for different apartment sizes, offices, or retail spaces. This morphology can support efficient land use, by reducing the percentage of built-area, in addition, natural ventilation and indoor daylighting can be favoured by the building shape and the presence of terraces. Finally, this typology can effectively supports energy efficiency and reduction of energy consumption (Saadatjoo et al. 2018).

Conversely, a two-blocks approach divides the required floor area into two separate buildings. This configuration provides more flexibility in orientation to optimize natural light and ventilation, creating a public or semi-private space between the buildings, which may serve as a garden, or shared access. In addition, it offers the possibility of different heights or functions for each block (e.g., one residential, one commercial), allowing for independent access points, which may improve circulation and emergency response (Steadman 2014).

The three-block design solution shares similar characteristics with the two-block design. In this case, green terraces are placed at different heights to enhance human-nature interaction, and green areas are designed according to buildings position. The building density in this solution is slightly lower than in the two previously described designs, resulting in a reduction of green open spaces.

Concerning the selection of NbS and related plant species, information provided by the previous Chaps. 3 and 4 and the site analysis constitute a starting point. Specifically, for the microclimate regulation analysis, different tree species are selected among to the list of occurrences, and their geometrical properties are considered for the subsequent modelling and performance simulations.

For the stormwater retention capacity, NbS such as green roofs, permeable pavements and swales and raingardens have been selected as the most performing solutions, and related herb species can be identified (Fig. 5.7). In particular, the selection of species for water-media NbS can be particularly challenging: specific species with low water needs and high resistance to temporary flooding has been selected, according to (Magliocco and et al. 2020), who defined guidelines for raingarden features in mediterranean urban environments.

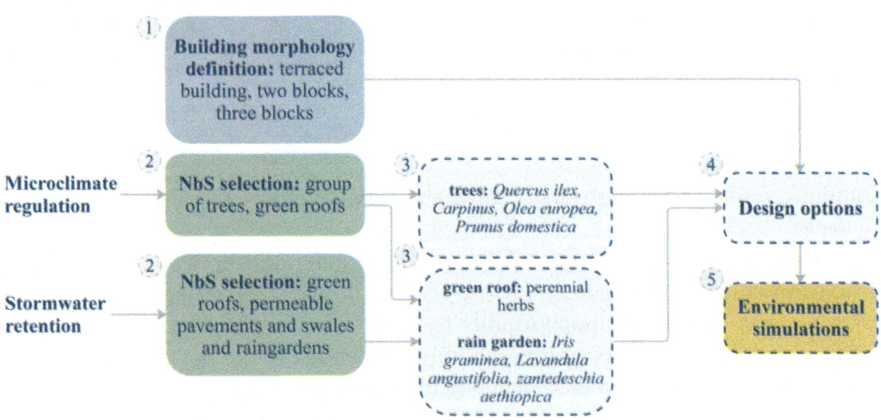

Fig. 5.7 Diagram on the selection of NbS and related species

5.2.4 *Environmental Simulations*

Starting from the 3D model of the site and the various building morphologies proposed in accordance with local requirements, described in the previous paragraph, microclimate and rainwater management simulations can be conducted. In particular, based on the results presented in Chaps. 3 and 4, the following tools have been selected: (1) ENVI-met integrated with Grasshopper for the microclimatic analysis, due to the higher level of detail it provides for NbS simulations, and (2) the native Grasshopper approach for stormwater management calculations, owing to its integration within the 3D CAD environment, which allows for an initial prediction of the performance of different NbS located on the selected site (Fig. 5.8).

Based on the climatic and rainfall data collected (see in Sect. 5.2.1), preliminary simulations can be setup:

(1) **Microclimatic analysis:** using the tools described in Chap. 3—specifically ENVI-met—the 3D geometry of the site can be converted into a format compatible with the CFD tool. This process begins with the definition of the grid, which determines the model's level of detail (in this case, 5 × 5 m). Once the grid is set, climatic data can be integrated, simulation parameters such as the selected day and duration can be configured, and the simulation can then be executed.

(2) **Stormwater analysis:** using the tools described in Chap. 4—particularly the Grasshopper-based approach—the 3D model can be further detailed to define specific characteristics of the various NbS, such as soil depth and drainage capacity. By introducing the precipitation rate for a selected time of year, simulations can then be carried out to evaluate different configurations and NbS scenarios.

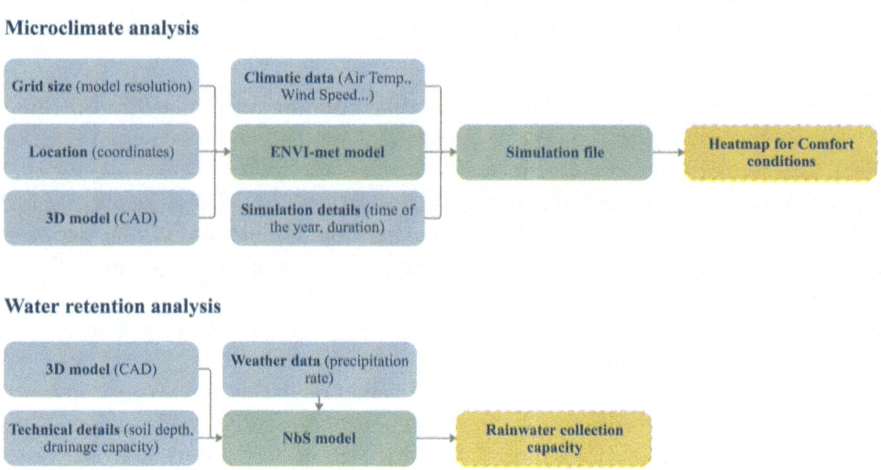

Fig. 5.8 Operative workflow for the environmental assessment of NbS performance in terms of microclimate regulation and rainwater collection

5.2.5 *Output Data*

The output data from the simulations are provided as .csv files for each design configuration considered and can also be visualized using heatmaps, particularly for microclimatic simulations. Heatmaps offer an overview and allow for the comparison of different design solutions. A more detailed data analysis can be conducted by examining the outputs (e.g., UTCI (°C), Air Temperature (°C)) for all points within the grid. Similarly, in water management simulations, colour gradients indicating the water retained by various nature-based solutions can be visualized for an easy comparison. The design solutions can also be compared in terms of overall performance, providing a clear indicator of rainwater collection capacity.

These data can be used in an optimization process to inform the selection and refinement of design solutions through feedback loops. The quantification of key performance indicators (KPIs), related to the design objectives identified during the design process, provides valuable insights for designing buildings that are adaptable to climate change and resilient to its impacts (Li et al. 2023; Farina et al. 2024).

According to the literature, the Multi-Objective Optimization (MOO) framework is used in urban planning to identify optimal strategies for selecting the most effective nature-based solutions for a specific site (Farina et al. 2024). Hybrid MCDM strategy are also applied to optimize design solutions at the building scale, such as integrating ecological knowledge into building design (Selvan et al. 2023a; Weisser et al. 2023), with significant benefits, enhancing the decision-making process (Selvan et al. 2023b).

5.3 Stormwater Retention and Thermal Comfort Performances and Optimization

In order to quantify the stormwater and thermal comfort performances of the three design solutions developed, specific models and simulation parameters are set up. First, the thermal comfort performance is calculated.

The selected tool for the microclimatic simulations, ENVI-met, requires a high level of detail for the quantification of thermal performance of both building geometry and vegetation. Table 5.3 shows the input parameters defined for the microclimate simulations.

The simulation results for the three solutions are summarized in Table 5.4 and related heatmap are shown in Fig. 5.9.

Figure 5.9 shows the distribution of UTCI values on the entire area and the contribution of buildings' shading, as well as the contribution of different surfaces (such as green areas with grass), and trees. In particular, in all the design scenarios the contribution of trees in wide open areas is particularly relevant.

Table 5.3 Model and simulation settings for microclimate performance assessment of the three design options

Buildings	Surfaces	Vegetation	Climatic data
Walls: concrete wall (lightweight) 0.02 roughness	Roads: Asphalt with gravel	Grass: grass 0.25 m average 0.2 albedo	21st June, 2024, timeframe 1–3PM, weather data from local .epw
Roofs: roofing tile 0.02 roughness		Trees: Olea europea (foliage albedo 0.5); Quercus ilex (foliage albedo 0.18); Carpinus (foliage albedo 0.18)	
Green envelope: green + sandy loam substrate			

Table 5.4 Microclimatic performance simulations on the three building morphologies and related output

Building morphology	Shaded UTCI (°C)	Sunlight UTCI (°C)	Shaded MRT (°C)	Sunlight MRT (°C)
Two blocks	27.15	33.96	32.58	57.31
Three blocks	27.78	34.73	34.27	62.02
Terraced building	29.47	34.74	43.69	61.27

In the two blocks scenario, UTCI values in open area can reach an average of 34.49 °C, while UTCI values where trees are located are reduced to 29.55 °C. Additionally, this configuration with two compact buildings with a higher amount of floors compared to the other options, provides a stronger shading on surrounding areas, reducing thermal stress conditions from 34.01 to 26.51 °C. The distribution of the two buildings on the site has also a strong impact on air flows: in particular, at the pedestrian level, the average wind speed (main wind direction from North to the South) on the site where the design is located is around 0.5 m/s. Around the two buildings, wind speed can reach 1.73 m/s, positively contributing to thermal comfort improvement.

The three blocks scenario is characterized by a similar thermal comfort improvement provided by the distribution of trees in outdoor spaces. Specifically, from the 34.59 °C of a sunlight cell, UTCI values are lowered by the shading and evaporation capacity of trees till 28.97 °C. Conversely, the shading contribution of the three buildings is lower than the previous options, due to the lower amount of floors for each block. Even if the lowest value of the thermal comfort index found in the immediate surrounding of the buildings is 26.76 °C (similar to the 26.51 °C provided by the previous option), the amount of shaded cells for this scenario is lower: specifically, the cells characterized by the shade provided by the three buildings are 27, while the amount of shaded cells in the two-blocks option is 38.

Finally, the terraced-building option presents an interesting solution, as it allows for a greater inclusion of vegetated surfaces on the building envelope. The building's geometry enables the creation of terraces and balconies that can accommodate more

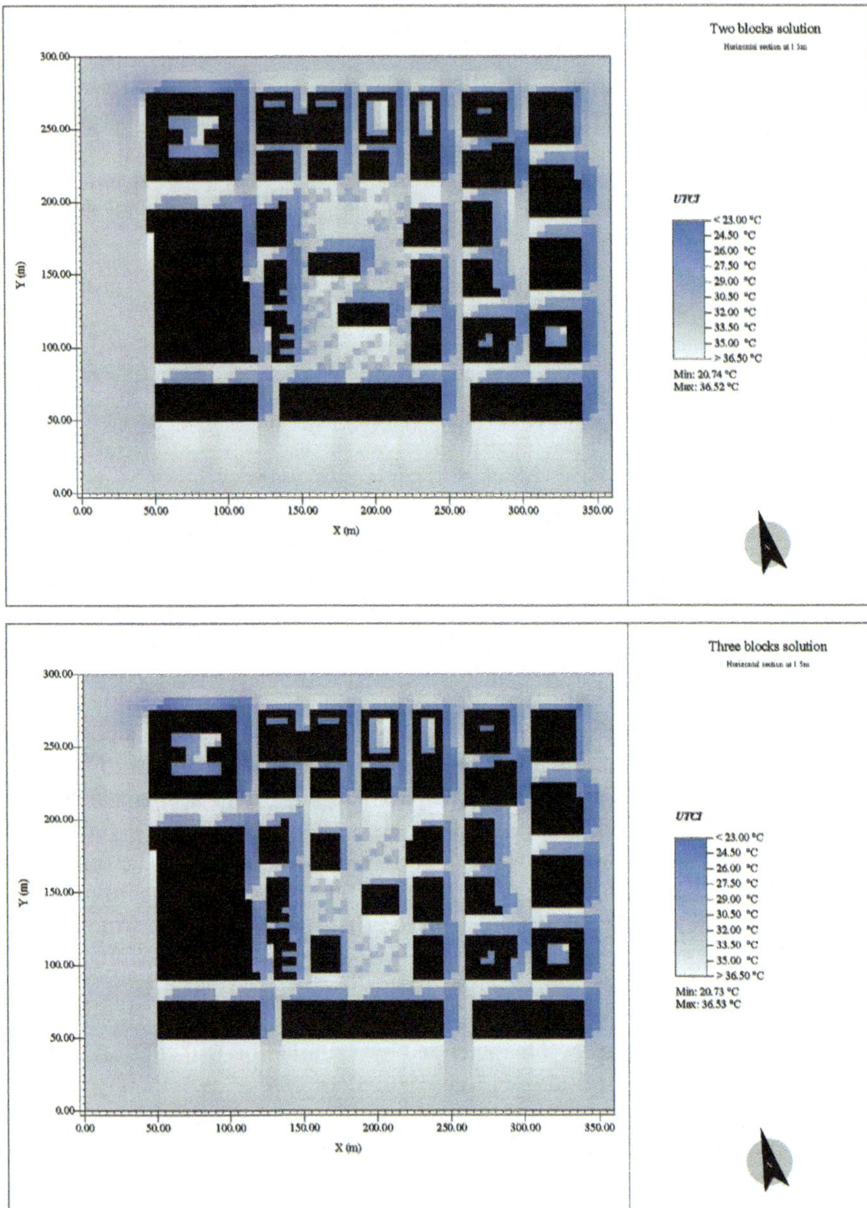

Fig. 5.9 Heatmap with the distribution of UTCI values for the three design options considered (two blocks, three blocks, and terraced building)

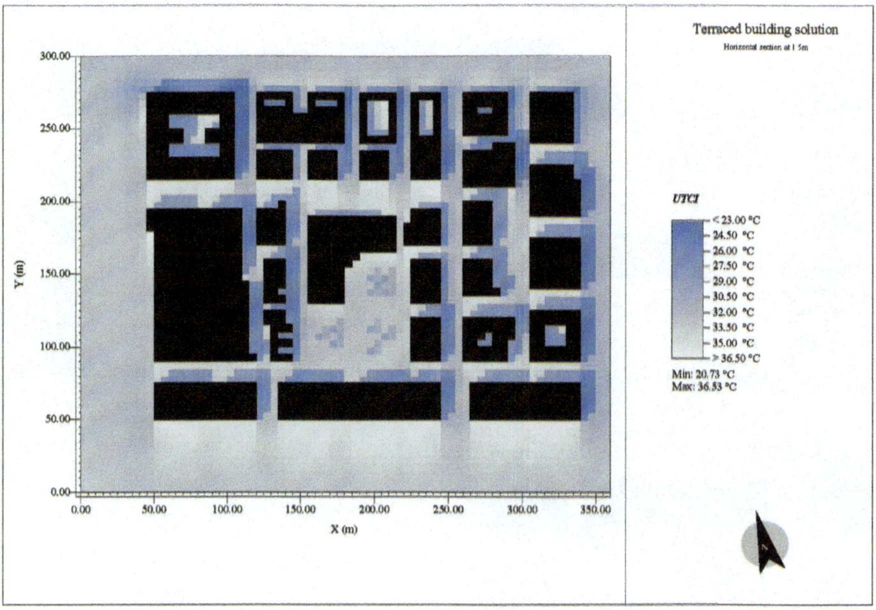

Fig. 5.9 (continued)

plants compared to the other two options. However, from a microclimatic perspective, the building shape results the less effective in terms of thermal stress reduction. Specifically, the tapered shape of the building and the related main distribution on the site at the ground floor does not provide a high amount of shaded areas in the surroundings of the building. However, the distribution of trees in open areas provides a relevant thermal comfort improvement, from 34.70 °C in sunlight areas to 28.76 °C. In addition, this configuration does not supports the optimization of air flows on the site, as the building constitutes a massive obstacle to the distribution of wind speed coming from North. Figure 5.10 shows a focus on the distribution of wind speed on the site for the three different design options.

Concerning the calculation of stormwater management, the selected methodology has been applied. In particular, the technical features of the different horizontal NbS on the three design scenarios have been defined, in order to allow a proper calculation of the performances of each scenario. Table 5.5 summarizes the characteristics of each NbS considered in the three scenarios.

The simulation results for the three design options and related NbS are summarized in Table 5.6. It is worth noting that the performance of each NbS depends not only on its technical features, such as the soil depth and the drainage capacity, but also on its surface extension. For this reason, the different design options, having a different ratio between open areas, pedestrian pathways, and roof surfaces, can obtain very different performance in terms of stormwater management. For example, the two blocks option has a reduced amount of surface dedicated to green roofs, however,

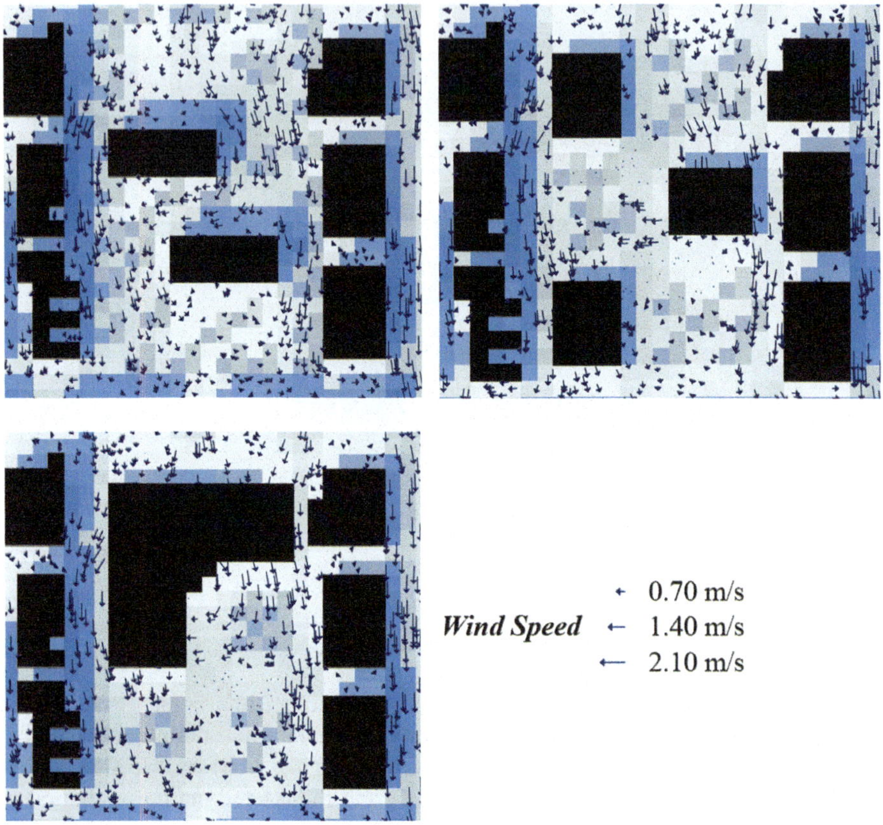

Fig. 5.10 Distribution of wind speed (m/s) on the site for the three different design options (two blocks, three blocks, and terraced building)

the surface for outdoor green areas is particularly favoured; conversely, the terraced building option provides an large amount of surface for roof greening for each floor, allowing the distribution of both extensive and intensive green roofs, but partially reducing the ground surface for greening.

A more detailed analysis of the results shows that the best performing solution in terms of stormwater management is the one with three blocks. Given the higher amount of permeable surfaces compared to the other two solutions, and the presence of intensive green roofs.

The percentage of coverage with green areas is lower, however, globally, the total amount of potential retention of rainwater during an extreme event is higher (675×10^3 m^3 of water, in total, compared to 527×10^3 m^3 for the two blocks solution and the 416×10^3 m^3 of the terraced building).

At this point, identifying a ratio that defines the amount of water retained by the different NbS independently from the extension of the surface is needed. By dividing the amount of water retainable by a NbS with the extension of the surface

Table 5.5 Model and simulation settings for stormwater management assessment of the three design options

Design options	NbS	Soil depth (m)	Downflow coefficient	Precipitation rate
Two blocks	Extensive green roof	0.1	0.15	260 mm/month (fall season)
	Permeable pavement	0.1	0.8	
	Green areas (grass)	0.4	0.3	
Three blocks	Intensive green roof	0.4	0.15	
	Permeable pavement	0.1	0.8	
	Green areas (grass)	0.4	0.3	
Terraced building	Intensive green roof	0.4	0.15	
	Extensive green roof	0.1	0.15	
	Green areas (grass)	0.4	0.3	

Table 5.6 Stormwater management performance simulations on the three building morphologies and related output

Design options	NbS	NbS total surface (m^2)	Water retained (m^3)
Two blocks	Extensive green roof	1353.00	20.3×10^3
	Permeable pavement	803.15	215.79×10^3
	Green areas (grass)	2425	291×10^3
Three blocks	Intensive green roof	1416.67	72×10^3
	Permeable pavement	1130.12	340×10^3
	Green areas (grass)	2192.22	263×10^3
Terraced building	Intensive green roof	1361.90	81.7×10^3
	Extensive green roof	1301.20	19.5×10^3
	Green areas (grass)	2625.9	315×10^3

of each solution, a ratio can be defined. Specifically, the following values have been calculated:

- Extensive green roofs: 15 m^3 of water/m^2 (max);
- Intensive green roofs: 60 m^3 of water/m^2 (max);
- Permeable pavements: 240 m^3 of water/m^2 (max);
- Green areas: 120 m^3 of water/m^2 (max).

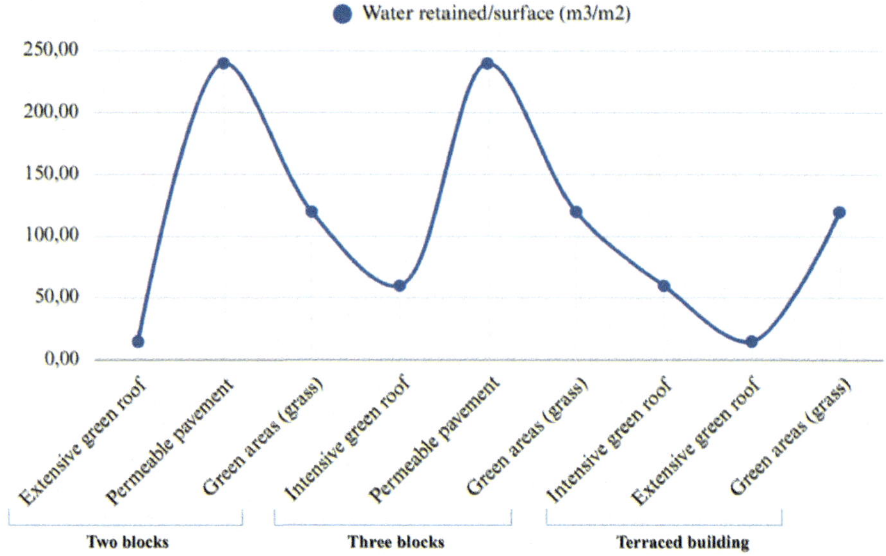

Fig. 5.11 Amount of rainwater potentially retained by the different NbS per square meter, for the three configurations modelled and simulated

Figure 5.11 shows the performance of the different NbS for each design scenario proposed for the site, allowing a comprehensive understanding of the different functioning in terms of stormwater management.

It is worth mentioning that the third design option (terraced building) focuses on maximizing the surface covered by intensive and extensive green roofs, in order to reduce the need for other types of NbS. However, the analysis of the stormwater management capacity has demonstrated the effectiveness of permeable and draining pavements, over all the other solutions proposed. By implementing this solution also in the terraced buildings option, covering 984 square meters of pedestrian paths, the overall capacity to retain and drain stormwater increases from 416 × 103 m^3 to 652 × 103 m^3, reaching the same performance of the three blocks solution.

Finally, starting from the results obtained, some considerations can be introduced in relation to the possibility to optimize the design solutions. As demonstrated in the paragraph above, through the simulation of NbS performance, given specific technical features of the different solutions, it is possible to define a ratio to allow their comparison. This ratio can be particularly useful in order to improve the performances of a specific design option, by manually modifying the amount of surfaces covered by green roofs, swales, green areas and drainage pavements.

The simulation of environmental performance of design options is indeed a method that can introduce iterative steps in the design process in order to improve the overall design quality (Mela et al. 2012). However, this iterative process can also be facilitated by the use of computational tools for multi-criteria optimization, in order to streamline the design process. Specifically, as already mentioned in Sect. 5.1, several multi-criteria methodologies can be implemented during the definition of a design solution. In relation to environmental performances of the design, Multi-objective optimization (MOO) methodologies and related tools are widely used (Paseka et al. 2018; Li et al. 2023; Mosca 2024). The main reason why these methodologies are widely implemented in environmental design is that a great variety of parameters must be considered in this context through the introduction of Key Performance Indicators. In particular, MOO algorithms primarily aim to achieve a balance between the performance of a design solution across different domains, minimizing trade-offs and maximizing synergies. In this case, optimized solutions meet the minimum performance thresholds established during the optimization criteria definition phase (Cui et al. 2017).

Concerning microclimate regulation and stormwater management capacity of a design solution, the implementation of optimization strategies is particularly important: for example, as demonstrated by the simulations, for thermal comfort improvement, the building shape and the position and number of trees are the most relevant parameters (Mosca et al. 2025), while for rainwater collection, the most influencing ones are the extension of horizontal surfaces with different soil depth and the drainage capacity (Kõiv-Vainik et al. 2022), and trees provide a limited performance compared to other NbS. Consequently, a design option with a relevant thermal comfort improvement not necessarily performing well in terms of stormwater management, as demonstrated by the application of the selected tools and related results.

5.4 Overview on Design and NbS Performances

In this section an overview of the performance of the different design configurations presented in this chapter is provided, in order to discuss the potentials and limitations of design options in terms of environmental performances (Fig. 5.12).

The **two blocks option** presents an optimal distribution of buildings that allow air flows in open areas, providing a positive contribution to thermal comfort improvement. It would be appropriate to verify the winter conditions in order to reduce risks of discomfort for high wind speed (Yin et al. 2021). The morphology of the buildings (tall and compact) improves outdoor thermal comfort thanks to the distribution of shadows in outdoor areas. In addition, the regular distribution of trees guarantees the reduction of mean radiant temperature, and the related improvement of microclimate conditions at the pedestrian level (Guo et al. 2020).

Concerning the stormwater management performance, the limited surface of roofs partially limits their capacity to retain water, however, the great availability of open areas at the ground level for green areas and pedestrian permeable paths guarantee an overall good performance. Given the high rate of green area, it would be advisable to propose the introduction of raingarden(s), in order to further improve the overall performance of the design option (US EPA 2015).

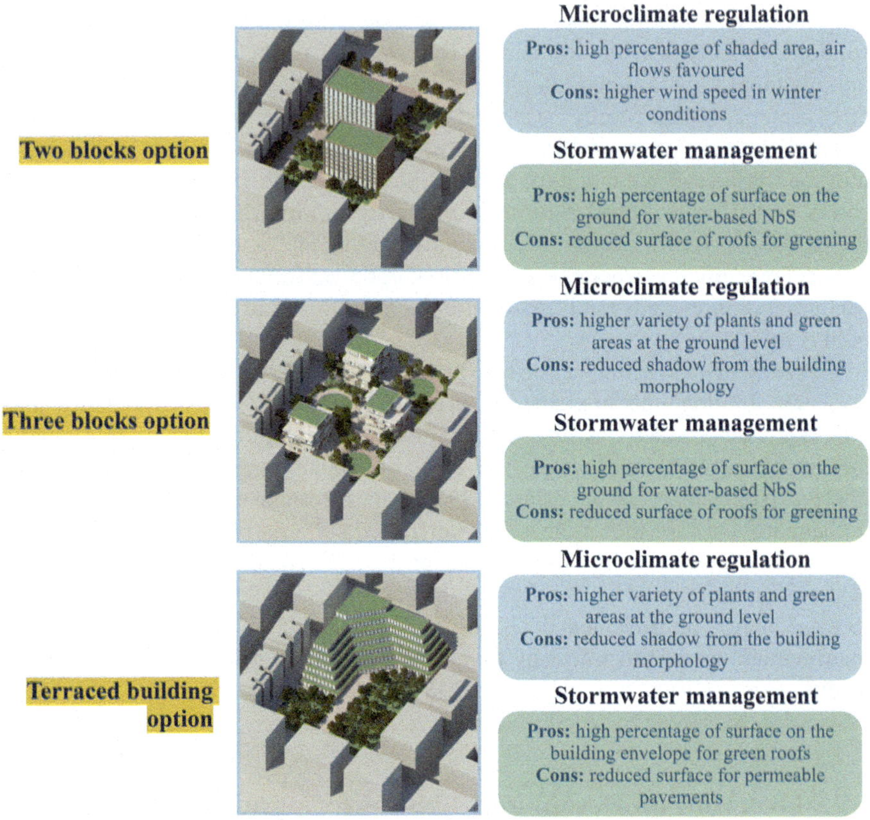

Fig. 5.12 Summary of potentials and limitations of each design solution for the two selected KPIs: microclimate regulation and stormwater retention

The **three blocks option** provides different performance compared to the previous solution: regarding microclimate improvement, this configuration is less effective, as it includes one more building, resulting in three blocks featuring lower heights. This aspect contributes to the reduction of shaded surfaces at the ground level, and, in addition, the wind analysis has demonstrated that the third building constitutes an "obstacle" for the air flows in the open area (Liu et al. 2022).

However, the performance of trees and ground-level green areas still provides a significant improvement in comfort conditions, comparable to the other solutions. Conversely, this configuration performs much better in terms of stormwater management, thanks to the greater availability of open ground areas for the implementation of NbS for rainwater retention and temporary storage. Results from this analysis demonstrate the relevant role of simulation and quantification tools for design performance and the importance of their implementation in the design process.

Finally, also the **terraced building option** performs differently from the previous solutions. In particular, in relation to microclimate regulation, the building configuration, characterized by a great development in horizontal, does not support the improvement of thermal comfort conditions. Indeed, the building morphology results in a reduced amount of shadow and poses limitations for air flows in open areas, resulting in UTCI values higher (2 °C more) compared to the two and three blocks' configurations. However, the role of open green areas and trees is consistent with the other options (Martini et al. 2020).

Concerning stormwater management, the high amount of surfaces for extensive and intensive green roofs contributes in collecting and retaining rainwater. As highlighted in Sect. 5.3, the overall performance of this solution is partially reduced compared to the others, as the ground surface exploitable for ground-based NbS is limited, and in the first design proposal, pedestrian paths were not designed as drainage pavements. After the first round of simulations, the crucial role of drainage surfaces has been demonstrated, and, by converting the existing pathways in permeable pavements, the overall performance of the design has increased.

It is worth noting that even if this design option results less effective in terms of environmental performances, it supports human-nature interaction, defined as one of the design objectives in Sect. 5.1. Specifically, the higher amount of terraces and green areas on the building envelope, in addition to the ground level green areas, represents an interesting starting point to improve human wellbeing (Bolten and Barbiero 2020) in terms of stress reduction and cultural ecosystem services provision (Babí Almenar et al. 2021).

5.5 Summary and Conclusions

This chapter provides a theoretical and operative perspective for the implementation of environmental simulations and analysis during the design process with NbS. The analysis of the case study has supported the illustration of the different steps proposed of the Nature-based Cities workflow. In particular, starting from the site analysis and data collection, simulation settings and the analysis of the outcomes is provided. The development of different design solutions on the selected site has supported the definition of a set of design recommendations for the implementation of NbS in urban environment for climate adaptation. In particular, the different features that define a design solution, such as the building shape, the spatial organization, and the definition of outdoor areas, can positively or negatively impact the overall performance both in terms of microclimate regulation and stormwater management. A key result is related to the building shape: for thermal comfort, the 3D volume of NbS (specifically trees) is crucial, while for stormwater retention the horizontal solutions and the soil depth are the most relevant parameters. It is important to note that the scale considered for the case study and the related simulations is relevant to the district level, but not to the city scale. The climate change adaptation strategies analyzed show significant performance in creating a cool spot, but they are not suitable as city-scale strategies for reducing the urban heat island effect. Similarly, to effectively manage stormwater, the planning and implementation of nature-based solutions should consider the city scale, using a treatment train approach (see Chap. 2).

References

Acquaotta F, Faccini F, Fratianni S et al (2018) Rainfall intensity in the Genoa Metropolitan Area: secular variations and consequences. Weather 73:356–362. https://doi.org/10.1002/wea.3208

Babí Almenar J, Elliot T, Rugani B et al (2021) Nexus between nature-based solutions, ecosystem services and urban challenges. Land Use Policy 100:104898. https://doi.org/10.1016/j.landusepol.2020.104898

Baggaley N, Britton A, Sandison F et al (2022) Understanding carbon sequestration from nature-based solutions. The James Hutton Institute

Barwise Y, Kumar P (2020) Designing vegetation barriers for urban air pollution abatement: a practical review for appropriate plant species selection. NPJ Clim Atmos Sci 3:1–19

Bolten B, Barbiero G (2020) Biophilic design: how to enhance physical and psychological health and wellbeing in our built environments. Vis Sustain 11–16. https://doi.org/10.13135/2384-8677/3829

Canepa M, Mosca F, Barath S et al (2022) Ecolopes, beyond greening. A multi-species approach for urban design. AGATHÓN Int J Arch Art Des 11:238–245. https://doi.org/10.19229/2464-9309/11212022

Cao X, Wang H, Zhang B et al (2024) Sustainable management of land use patterns and water allocation for coordinated multidimensional development. J Clean Prod 457:142412. https://doi.org/10.1016/j.jclepro.2024.142412

Chang K-H (2015) Chapter 19—Multiobjective optimization and advanced topics. In: Chang K-H (ed) E-design. Academic Press, Boston, pp 1105–1173

Chen S-J, Hwang C-L (1992) Fuzzy multiple attribute decision making methods. In: Chen S-J, Hwang C-L (eds) Fuzzy multiple attribute decision making: methods and applications. Springer, Berlin, Heidelberg, pp 289–486

Cui Y, Geng Z, Zhu Q, Han Y (2017) Review: multi-objective optimization methods and application in energy saving. Energy 125:681–704. https://doi.org/10.1016/j.energy.2017.02.174

Dale B, Dananto M, Kifle B (2025) Dynamics of land use land cover change and its effect on urban heat island in Halaba Kulito Town. Heliyon 11:e41689. https://doi.org/10.1016/j.heliyon.2025.e41689

Directorate-General for Research and Innovation (European Commission), Dumitru A, Wendling L (2021) Evaluating the impact of nature-based solutions: appendix of methods. Publications Office of the European Union

Eyring V, Bony S, Meehl GA et al (2016) Overview of the coupled model intercomparison project phase 6 (CMIP6) experimental design and organization. Geosci Model Dev 9:1937–1958. https://doi.org/10.5194/gmd-9-1937-2016

Farina G, Neverre N, Hérivaux C et al (2024) How to account for spatial trade-offs in planning for urban climate adaptation? Optimizing green and grey infrastructures. J Environ Manage 372:123380. https://doi.org/10.1016/j.jenvman.2024.123380

Federal Interagency Stream Restoration Working Group (1998) Stream corridor restoration: principles, processes, and practices: GPO Item No. 0120-A; SuDocs No. A 57.6/2:EN 3/PT.653

Ferreira O, Viavattene C, Jiménez JA et al (2018) Storm-induced risk assessment: evaluation of two tools at the regional and hotspot scale. Coast Eng 134:241–253. https://doi.org/10.1016/j.coastaleng.2017.10.005

Global Economic Forum (2024) The global risks report 2024, 19th edn

Guo H, Aviv D, Loyola M et al (2020) On the understanding of the mean radiant temperature within both the indoor and outdoor environment, a critical review. Renew Sustain Energy Rev 117:109207. https://doi.org/10.1016/j.rser.2019.06.014

Hwang C-L, Yoon K (1981a) Methods for multiple attribute decision making. In: Hwang C-L, Yoon K (eds) Multiple attribute decision making: methods and applications a state-of-the-art survey. Springer, Berlin, Heidelberg, pp 58–191

Hwang C-L, Yoon K (1981b) Multiple attribute decision making. Springer, Berlin, Heidelberg

GBIF.Org User (2025) Occurrence Download. 20261

Intergovernmental Panel On Climate Change (IPCC) (2023) Climate Change 2022—impacts, adaptation and vulnerability: working group II contribution to the sixth assessment report of the intergovernmental panel on climate change, 1st edn. Cambridge University Press

Kazimieras Zavadskas E, Antucheviciene J, Kar S (2019) Multi-objective and multi-attribute optimization for sustainable development decision aiding. Sustainability 11:3069. https://doi.org/10.3390/su11113069

Kõiv-Vainik M, Kill K, Espenberg M et al (2022) Urban stormwater retention capacity of nature-based solutions at different climatic conditions. Nature Based Solut 2:100038. https://doi.org/10.1016/j.nbsj.2022.100038

Li Z, Zhao Y, Xia H, Xie S (2023) A multi-objective optimization framework for building performance under climate change. J Build Eng 80:107978. https://doi.org/10.1016/j.jobe.2023.107978

Liu K, Lian Z, Dai X, Lai D (2022) Comparing the effects of sun and wind on outdoor thermal comfort: a case study based on longitudinal subject tests in cold climate region. Sci Total Environ 825:154009. https://doi.org/10.1016/j.scitotenv.2022.154009

Ma K-T, Luo Y, Kwan T, Wu Y (2019) Chapter 3—Environmental loads and vessel motions. In: Ma K-T, Luo Y, Kwan T, Wu Y (eds) Mooring system engineering for offshore structures. Gulf Professional Publishing, pp 41–62

Magliocco A et al (2020) Infrastrutture Verdi Per L'adattamento AI Cambiamenti Climatici Strategie e indicazioni progettuali per la gestione sostenibile delle acque meteoriche urbane nell'area mediterranea nord-occidentale

Martini A, Biondi D, Batista AC (2020) Thermal comfort provided by street trees in cities. Arboricult J 42:153–164

Mela K, Tiainen T, Heinisuo M (2012) Comparative study of multiple criteria decision making methods for building design. Adv Eng Inform 26:716–726. https://doi.org/10.1016/J.AEI.2012.03.001

Mosca F (2024) A conceptual framework for the optimization of environmentally sustainable nature-based solutions. Wichmann Verlag, DE

Mosca F, Calbi M, Roccotiello E, Perini K (2025) A computational approach to assess the effects of ecological building envelopes on outdoor thermal comfort. Sustain Cities Soc 120:106170. https://doi.org/10.1016/j.scs.2025.106170

Paseka S, Kapelan Z, Marton D (2018) Multi-objective optimization of resilient design of the multipurpose reservoir in conditions of uncertain climate change. Water 10:1110. https://doi.org/10.3390/w10091110

Saadatjoo P, Mahdavinejad M, Zhang G (2018) A study on terraced apartments and their natural ventilation performance in hot and humid regions. Build Simul 11:359–372. https://doi.org/10.1007/s12273-017-0407-7

Selvan SU, Saroglou ST, Joschinski J et al (2023b) Toward multi-species building envelopes: a critical literature review of multi-criteria decision-making for design support. Build Environ 231:110006. https://doi.org/10.1016/j.buildenv.2023.110006

Selvan S, Mosca F, Perini K et al (2024) Resilient green building envelopes: a computational method for holistic sustainability assessments and interdisciplinary design decision-making. Nicosia, pp 403–412

Selvan SS, Mosca F et al (2023a) Multi-species building envelopes: developing a multi-criteria decision-making methodology for cohabitation. India

Sharma R, Malaviya P (2021) Management of stormwater pollution using green infrastructure: the role of rain gardens. Wires Water 8:e1507. https://doi.org/10.1002/wat2.1507

Steadman P (2014) Building types and built forms. Troubador Publishing Ltd

Taha H (1997) Urban climates and heat islands: albedo, evapotranspiration, and anthropogenic heat. Energy Build 25:99–103. https://doi.org/10.1016/S0378-7788(96)00999-1

US EPA R 01 (2015) Soak up the rain: rain gardens. https://www.epa.gov/soakuptherain/soak-rain-rain-gardens. Accessed 6 Sep 2024

Vinogradova I (2019) Multi-attribute decision-making methods as a part of mathematical optimization. Mathematics 7:915. https://doi.org/10.3390/math7100915

Weerasuriya AU, Zhang X, Wang J et al (2021) Performance evaluation of population-based meta-heuristic algorithms and decision-making for multi-objective optimization of building design. Build Environ 198:107855. https://doi.org/10.1016/j.buildenv.2021.107855

Weisser WW, Hensel M, Barath S et al (2023) Creating ecologically sound buildings by integrating ecology, architecture and computational design. People Nat 5:4–20. https://doi.org/10.1002/pan3.10411

Yin Q, Cao Y, Sun C (2021) Research on outdoor thermal comfort of high-density urban center in severe cold area. Build Environ 200:107938. https://doi.org/10.1016/j.buildenv.2021.107938

Chapter 6
Design Recommendations for Nature-Based Cities

Abstract This Chapter presents design recommendations for Nature-based Cities, considering the primary challenges urban areas are currently facing, both in terms of environmental and ecological issues. Building on the approaches and research results presented throughout the book, it links design strategies at different scales with the impacts of climate change (CC) on cities and their inhabitants to address the increasing complexity of urban environments. Results from the case study applications provide the foundation for defining an overview of the most relevant design features for climate adaptation, highlighting the trade-offs and synergies between solutions for microclimate regulation and stormwater management. Finally, overall conclusions on the role of nature-based solutions in cities to enhance urban resilience are presented.

Keywords Nature-based solutions · Computational design · Climate change adaptation · Microclimate regulation · Water management · Adaptation strategies · Climate change impacts

6.1 Facing the Challenges of Urbanization

The report on Nature-based Solutions for Comprehensive Disaster and Climate Risk Management (UNDRR 2024) developed by the United Nations Office for Disaster Risk Reduction (UNDRR) and the United Nations University (UNU) outlines key factors to consider when planning and implementing nature-based solutions through an integrated approach. Figure 6.1 highlights the relationship between these factors, which are related to environment, climate, hazards, exposure, vulnerability and impacts. In this Chapter, the focus is on the types of analyses and data to be collected at the city and neighbourhood scale as the initial step in the planning and implementation process.

K. Perini and F. Mosca, *Nature-Based Cities*, Digital Innovations in Architecture, Engineering and Construction, https://doi.org/10.1007/978-3-031-94612-7_6

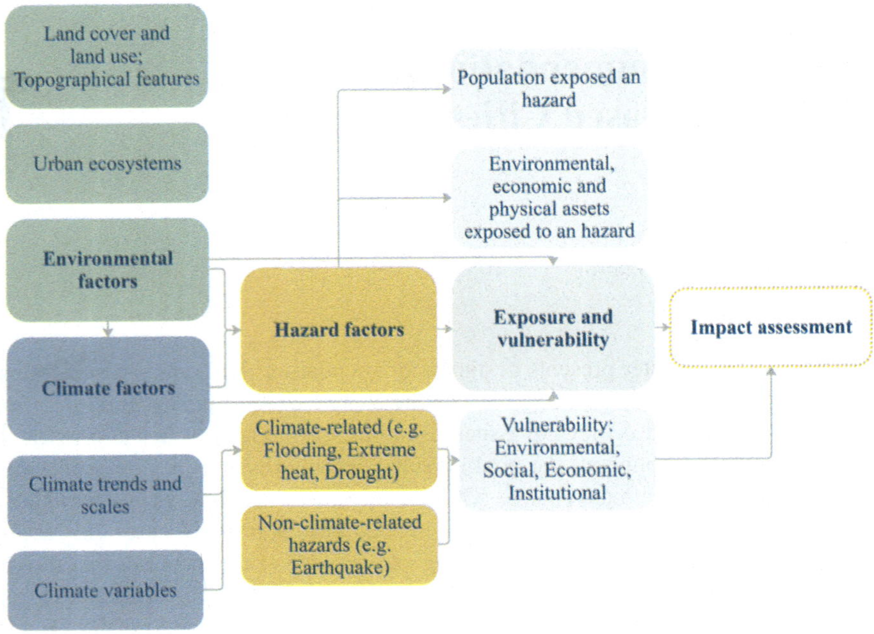

Fig. 6.1 Connection between environment, climate, hazards, exposure, vulnerability and impacts main factors to ponder before considering nature-based solutions options, based on UNDRR and UNU-EHS (2024)

Starting from the environmental factors, land use and land cover include, for example, settlements, infrastructure, water bodies and watersheds; examples of topographical features are elevation and slope of a site. Different types of urban ecosystems, such as urban parks and forests, urban gardens and agriculture, green roofs/walls/sidewalks and other urban features, urban water bodies, sustainable drainage systems and urban nature reserves (Farr 2011), can be identified (Rees 1997). Land cover or use can be analysed using different approaches, such as computational methods (Zhang et al. 2023), relining, for example on opensource databases as Copernicus, and planning tools provided at the national or local level. The selection of the best approach, computational or analogue, depends on factors as site complexity, available information, and requirements in terms of detail, among others.

The identification of the most relevant urban ecosystems can also rely on the approaches mentioned above (Li et al. 2019) or on the use of open-source databases, such as the IUCN Global Ecosystem Typology (IUCN 2020), which includes a comprehensive classification framework for Earth's ecosystems. An analysis of city-level plans can also provide useful information on existing ecosystems at a smaller scale, helping to identify those that can be strengthened and/or those that need to be implemented to ensure the provision of relevant ecosystem services.

As outlined in Chap. 5, climate data provide planning and design-relevant information and are necessary inputs for environmental simulations. Collecting climate and climate change data at the city scale allows for identifying trends and scenarios linked to climate-related hazards (Demuzere et al. 2022).

Hazards can be both climate-related and non-climate related. In the first case, nature-based solutions can act on both causes and effects, linking mitigation and adaptation strategies to increase urban resilience (Meerow et al. 2016; Frantzeskaki et al. 2022). Differently, non-climate related hazards as earthquakes can be considered in the planning process of cities as relevant design constrains, to be faced according to local regulations.

Exposure and vulnerability are related to social, economic, and environmental factors (UNDRR 2024). More specifically, exposure includes both population groups and environmental, economic, and physical assets exposed to a certain hazard. The environmental dimension of vulnerability is related to environmental quality, the social dimension to demographic patterns, inequality and marginalization, health and well-being, etc., the economic dimension to livelihoods and financial buffer capacities. Finally, the institutional dimension is also relevant to the planning and implementation of nature-based solutions (NbS), as it considers, for example, governance and institutional capacities. Impact assessment allows evaluating the increase in risks and vulnerabilities as a result of a hazard event.

At the planning level, a holistic analysis allows for linking environmental, climate, and hazard factors to exposure and vulnerability. Exposure considers the population groups, environmental, economic, and physical assets that are exposed to a particular hazard (Yu et al. 2021). Vulnerability addresses the environmental, social, economic, and institutional dimensions (Romero Lankao and Qin 2011). When planning and designing nature-based solutions, it is possible to identify the most effective sites or areas to, for example, improve health and well-being, enhance livelihoods, and reduce inequity and marginalization (Croeser et al. 2021).

Classifications of urban landscapes can facilitate the acquisition of interdisciplinary knowledge, considering also ecological-relevant data. For example, Oneto et al. 2024 represent the layered urban dimensions through four different subclassifications: local and landscape Urban Form, Anthropic Imprint, and Biophysical Conditions.

In the design and planning process, such level of analysis can be used to: (a) inform the selection of a site where to design nature-based solutions, (b) facilitate an holistic and comprehensive data collection for a specific site. In the design and planning process, this level of analysis and data can be used to: (a) inform the selection of a site for designing nature-based solutions, and (b) facilitate a holistic and comprehensive data collection for a specific site.

Such classification and analysis approaches are not new (McHarg 1969), but today, they integrate high-complexity, georeferenced data and current and future scenarios that can be incorporated into the design process.

The UNDRR (2024) highlights the importance of identifying NbS and NbS-related statements/concepts in existing plans, mainly in the national policy and planning landscape in relation to Disaster Risk Reduction (DRR) and Climate Change Adaptation (CCA), as a preparatory step for an integrated planning process. According to the report, the next step is the selection of the type of NbS to address site-specific challenges (which can be identified starting from the factors shown in Fig. 6.1). According to this approach, NbS are first selected based on climate-related hazards, such as floods or heatwaves.

The selection of NbS can rely on existing international databases, that group these solutions based on different features (Eisenberg et al. 2019; Somarakis et al. 2019; European Commission. Directorate General for Research and Innovation 2021): dimensions and scale for implementation (building, district, urban scale), type of performance (rainwater collection, particulate matter absorption, microclimate regulation), or type of natural elements included (trees and shrubs, water-related, etc.).

In the planning and implementation of NbS, stakeholder engagement plays a key role. As broadly discussed, for example, in the processes for river restoration, both bottom-up and top-down approaches to regenerative design rely on the involvement of local communities, as well as local and national governments (Perini and Sabbion 2017).

6.2 Lessons Learned from Performance-Driven Design Approach

As outlined in Sect. 6.1, the use of data to support the implementation of CCA and risk reduction strategies is crucial. Building on the broad description of factors presented in Fig. 6.1, and informed by the case study application in Chap. 5, the topic can be narrowed down for the identification of specific data packages to be used during the design process, with a focus on stormwater retention and thermal comfort.

First, the urban morphology plays a relevant role. Through the site analysis with Digital Elevation Models (DEM) and satellite data, a comprehensive understanding of the site can be achieved (topographical features). Climatic data collection from local weather stations and weather files (.epw) allows for the characterization of the site from an environmental perspective and supports the simulation of the impact of design options on local conditions.

Based on the information collected, an in-depth analysis of potential climate related risks on the site can be performed, and meaningful data can be produced to inform the design process and improve the overall design performances.

Starting from this approach, further steps could be developed and integrated in the design process in order to predict and quantify not only the effects of adaptation strategies on environmental conditions, but also the potential reduction of risks and extreme events in the long-term perspective (see Chap. 1, Sect. 1.2).

Finally, the application of the Nature-based Cities conceptual and computational approach has also allowed the identification of the most relevant and influential design parameters that should be considered during the design process for performance optimization. Three main categories of features can be identified: building-related features, trees-related features, and water-based NbS-related features. Below, an overview of these features is provided.

First, **building morphology** has a strong impact on the environmental quality of the entire site. Specifically, taller and more compact building shapes can create more shade on the ground, providing better thermal comfort at pedestrian level. At the same time, they offer more space at ground level for water-based NbS and strategies for stormwater management; additionally, the **building's location** on the site, depending on the morphology and configuration of the area, is particularly relevant. Indeed, the position of the building can be optimized to enhance airflows during summer conditions in open areas, improving perceived thermal comfort for people. On the other hand, during winter, the increased wind speed can cause discomfort conditions and reduce the liveability of outdoor spaces (Perini et al. 2018): an in-depth analysis of the performance in both seasons should be then performed.

The second category is related to land-based NbS: in particular, the position of **trees and shrubs**, as well as their geometrical configurations in outdoor spaces, is crucial for improving microclimate conditions. Regarding this aspect, the main focus during the design process should be on the distribution of airflows and shading on the ground. From an environmental and ecological perspective, selecting the right species is also important to enhance the overall resilience of the local ecosystem.

The third and last category is related to water-based NbS: specifically, the **depth of soil and the drainage capacity** of the different solutions are the most impactful parameters. The selection of the materials is then crucial, and their optimization should be favoured.

A more detailed analysis of the six parameters listed, and the related potentials and limitations, allows the identification of mutual relationships: for example, the building morphology has a direct connection with the technical features of water-based NbS: in particular, depending on the availability of surface for roof and terraces greening, different strategies can be implemented (e.g. extensive or intensive green roof technology), and the related performance in terms of rainwater collection can be limited or promoted (Palla and Gnecco 2018; Longobardi et al. 2019). In addition, soil depth and drainage capacity have also a strong impact on building's structural integrity, and related overall morphology (Susca 2019). Similarly, the position of the building on the site has a direct connection with the position of the selected plants, that should be optimized in order to improve the amount of shaded surfaces (Antoniadis et al. 2018; Martini et al. 2020). The selection of plant species, based on the defined configuration and the environmental conditions (soil depth), can also perform a relevant ecological performance in terms of biodiversity improvement and conservation (Mayrand and Clergeau 2018), and supports the retention of rainwater (Berland et al. 2017).

Figure 6.2 summarizes the parameters identified and their impact on microclimate regulation and stormwater management, highlighting the potentials and limitations of each feature described.

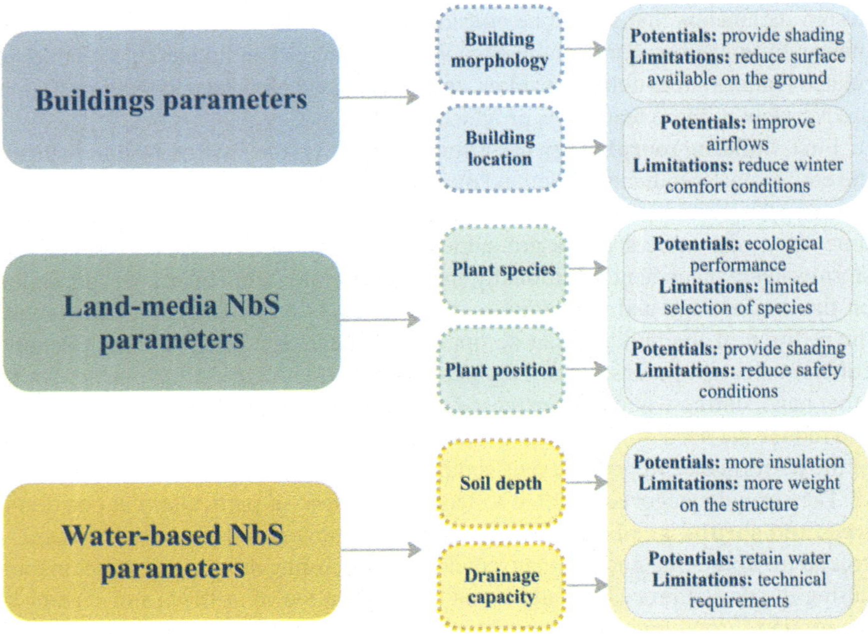

Fig. 6.2 Diagram summarizing the most relevant parameters and factors to be considered during the design process to optimize the environmental performances

6.3 Summary and Conclusions

The main aim of Nature-based Cities was to provide a comprehensive overview of the role of nature-based solutions in urban areas, in relation to environmental benefits, with a focus on the capacity to improve local adaptation to climate change. By describing the role of anthropic activities and related emissions, the most relevant environmental and climatic challenges are described (Chap. 1). Nature-based solutions are then introduced as a set of solutions with a high level of versatility for their implementation, inspired by natural processes, that can support the improvement of climate resilience in urban areas. The most relevant performances of NbS, and related indicators for their quantification, have been illustrated and described in detail (Chap. 2). Microclimate regulation and stormwater management were selected as the focus of the book, as they are among the most critical challenges cities are currently facing. Design solutions should not only consider current challenges, but also future perspectives in relation to climate change effects, resulting in more frequent and intense extreme events. An in-depth analysis of the currently available computational tools for the two selected performance areas (Chaps. 3 and 4) allowed for the identification of relevant design and technical parameters to be considered during the design process and provided an overview of the potentials and limitations of simulation tools for NbS performance quantification. Two methodologies were finally selected as the most suitable and versatile for their integration in a computational design process. The application of the Nature-based Cities design process and the related workflow to the selected case study has highlighted the importance and versatility of computational tools in supporting the decision-making process for designing adaptable urban districts. On the other hand, the role of designers in defining overall design objectives and evaluating different design options has been underlined (Chap. 5).

A set of design recommendations has been developed based on the discussion of the case study results, allowing a detailed and comprehensive understanding of NbS behavior under extreme weather conditions. These considerations constitute a starting point for the definition of guidelines for NbS implementation at the district level. Additional KPIs could be included in the design process of Nature-based Cities depending on site-specific requirements and critical aspects. The operative framework proposed is flexible and could be adapted based on designers and design needs.

Finally, it is worth mentioning the importance of trade-offs and synergies of the design solutions in relation to different environmental aspects considered. The application of the Nature-based Cities approach in the design of urban areas at different scales constitutes a starting point for more resilient, adaptable, and ecological cities.

References

Antoniadis D, Katsoulas N, Kittas C (2018) Simulation of schoolyard's microclimate and human thermal comfort under Mediterranean climate conditions: effects of trees and green structures. Int J Biometeorol 62:2025–2036. https://doi.org/10.1007/s00484-018-1612-5

Berland A, Shiflett SA, Shuster WD et al (2017) The role of trees in urban stormwater management. Landsc Urban Plan 162:167–177. https://doi.org/10.1016/j.landurbplan.2017.02.017

Croeser T, Garrard G, Sharma R et al (2021) Choosing the right nature-based solutions to meet diverse urban challenges. Urban Urban Green 65:127337. https://doi.org/10.1016/j.ufug.2021.127337

Demuzere M, Kittner J, Martilli A et al (2022) A global map of local climate zones to support earth system modelling and urban-scale environmental science. Earth Syst Sci Data 14:3835–3873. https://doi.org/10.5194/essd-14-3835-2022

Eisenberg B, Chiesa C, Fischer LK et al (2019) Nature-based solutions technical handbook factsheets

European Commission. Directorate General for Research and Innovation (2021) Evaluating the impact of nature-based solutions: a handbook for practitioners. Publications Office, LU

Farr D (2011) Sustainable urbanism: urban design with nature, 1st edn. Wiley

Frantzeskaki N, Ossola A, Bush J (2022) Nature-based solutions for changing urban landscapes: Lessons from Australia. Urban for Urban Green 73:127611. https://doi.org/10.1016/j.ufug.2022.127611

IUCN (2020) IUCN global ecosystem typology. https://iucn.org/resources/conservation-tool/iucn-global-ecosystem-typology. Accessed 22 Mar 2025

Li E, Parker SS, Pauly GB et al (2019) An urban biodiversity assessment framework that combines an urban habitat classification scheme and citizen science data. Front Ecol Evol 7. https://doi.org/10.3389/fevo.2019.00277

Longobardi A, D'Ambrosio R, Mobilia M (2019) Predicting stormwater retention capacity of green roofs: an experimental study of the roles of climate, substrate soil moisture, and drainage layer properties. Sustainability 11:6956. https://doi.org/10.3390/su11246956

Martini A, Biondi D, Batista AC (2020) Thermal comfort provided by street trees in cities. Arboricult J 42:153–164

Mayrand F, Clergeau P (2018) Green roofs and green walls for biodiversity conservation: a contribution to urban connectivity? Sustainability 10:985. https://doi.org/10.3390/su10040985

McHarg IL (1969) Design with nature. https://www.ibs.it/design-with-nature-libro-inglese-ian-l-mcharg/e/9780471114604. Accessed 9 Nov 2023

Meerow S, Newell JP, Stults M (2016) Defining urban resilience: a review. Landsc Urban Plan 147:38–49. https://doi.org/10.1016/j.landurbplan.2015.11.011

Oneto G, Perini K, Canepa M et al (2024) Urban cohabscapes: exploring European co-habitative landscapes diversity, in the ECOLOPES framework

Palla A, Gnecco I (2018) Green roofs to improve water management. In: Nature based strategies for urban and building sustainability. Elsevier, pp 203–213

Perini K, Sabbion P (2017) Urban sustainability and river restoration: green and blue infrastructure. Wiley

Perini K, Chokhachian A, Auer T (2018) Chapter 3.3—Green streets to enhance outdoor comfort. In: Pérez G, Perini K (eds) Nature based strategies for urban and building sustainability. Butterworth-Heinemann (Elsevier), Oxford, United Kingdom, pp 119–129

Rees WE (1997) Urban ecosystems: the human dimension. Urban Ecosyst 1:63–75. https://doi.org/10.1023/A:1014380105620

Romero Lankao P, Qin H (2011) Conceptualizing urban vulnerability to global climate and environmental change. Curr Opin Environ Sustain 3:142–149. https://doi.org/10.1016/j.cosust.2010.12.016

Somarakis G, Stagakis S, Chrysoulakis N (2019) ThinkNature/nature-based solutions handbook. https://doi.org/10.26225/JERV-W202

Susca T (2019) Green roofs to reduce building energy use? A review on key structural factors of green roofs and their effects on urban climate. Build Environ 162:106273

UNDRR (2024) Nature-based solutions for comprehensive disaster and climate risk management: Toolkit for integrated planning and implementation of disaster risk reduction and climate change adaptation. https://www.undrr.org/publication/nature-based-solutions-comprehensive-disaster-and-climate-risk-management-toolkit. Accessed 22 Mar 2025

Yu J, Castellani K, Forysinski K et al (2021) Geospatial indicators of exposure, sensitivity, and adaptive capacity to assess neighbourhood variation in vulnerability to climate change-related health hazards. Environ Health 20:31. https://doi.org/10.1186/s12940-021-00708-z

Zhang P, Ghosh D, Park S (2023) Spatial measures and methods in sustainable urban morphology: a systematic review. Landsc Urban Plan 237:104776. https://doi.org/10.1016/j.landurbplan.2023.104776

The manufacturer's authorised representative in the EU is Springer Nature Customer Service Centre GmbH, Europaplatz 3, 69115 Heidelberg, Germany. If you have any concerns regarding our products, please contact ProductSafety@springernature.com

Printed and bound by CPI Group (UK) Ltd, Croydon, CR0 4YY
30/09/2026
02225469-0002